Fermentation

History of Science and Medicine Library

VOLUME I

Fermentation

Vital or Chemical Process?

by

Joseph S. Fruton

BRILL

BRILL
LEIDEN • BOSTON
2006

This book is printed on acid-free paper.

Library of Congress Cataloging-in-Publication data
A C.I.P. record for this book is available from the Library of Congress.

ISSN 1872-0684
ISBN 90 04 15268 7
ISBN 978 90 04 15268 7

PRINTED IN THE NETHERLANDS

To the memory of
FREDERIC LAWRENCE HOLMES
(1932–2003)

CONTENTS

ACKNOWLEDGEMENTS

I am greatly indebted to O. Theodore Benfey, Jerome A. Berson, Roald Hoffmann, and Alan J. Rocke for their valuable criticisms and suggestions for an early draft of this book, and to Ramona Moore for her warmhearted assistance.

INTRODUCTION

Nobody who ever lived, now living, or to come,
will understand correctly the nature of fermentation

(J. Kunckel, 1630–1703)[1]

Shortly before he died, Isaac Newton (1642–1727) deposited with his printer the manuscript of the fourth edition of his *Opticks*, published in 1730. In the famous Query 31, beloved of historians of chemistry, he wrote:

> . . . we may learn that sulphureous Streams abound in the Bowels of the Earth and ferment with Minerals, and sometimes take fire with a sudden Coruscation and Explosion . . . Also some sulphureous Streams, at all times when the Earth is dry, ascending into the Air, ferment there with nitrous Acids, sometimes taking fire cause Lightning and Thunder, and fiery meteors. For the Air abounds with acid Vapours fit to promote Fermentations, as appears by the rusting of Iron and Copper in it, the kindling of fire by blowing, and the beating of the Heart by means of Respiration. Now the above-mention'd Motions are so great and violent as to shew that in Fermentations the Particles of Bodies which almost rest, are put into new Motions by a very potent Principle, which acts upon them only when they approach one another, and causes them to meet and clash with great violence, and grow hot with the motion, and dash one another into pieces, and vanish into Air, and Vapour, and Flame.[2]
>
> All these things being consider'd, it seems probable to me, that God in the Beginning form'd Matter in solid, massy, hard, impenetrable, moveable Particles . . . It seems to me further, that these Particles have not only a *Vis inertiae*, accompanied with such passive Laws of Motion as naturally result from that Force, but also that they are moved by certain active Principles, such as is that of Gravity, and that which causes Fermentation, and the Cohesion of Bodies. These Principles I consider, not as occult Qualities, supposed to result from the specifick Forms of Things, but as general Laws of Nature, by which the Things themselves are form'd; their Truth appearing to us by Phaenomena,

---

[1] Kunckel, J. (1716). *Collegium Physico-Chymicum Experimentale oder Laboratorium Chymicum*, p. 697. Hamburg: éHeyl. Facsimile reprint (1975). Hildesheim: Georg Olms.

[2] Newton, I. (1730). *Opticks*, pp. 379–380. Fourth ed. (Facsimile edition, 1931). London: Bell.

> though their Causes be not yet discover'd. For these are manifest Qualities, and their Causes only are occult. And the *Aristotelians* gave the Name of occult Qualities, not to manifest Qualities, but to such Qualities only as they supposed to lie hid in Bodies, and to be the unknown Causes of manifest Effects: Such as would be the Causes of Gravity, and of magnetick and electrick Attractions, and of Fermentations, if we should suppose that these Forces or Actions arose from Qualities unknown to us, and uncapable of being discovered and made manifest.[3]

Such statements about "fermentations" involving "sulphureous streams" and "nitrous acids" were not uncommon in England during 1650–1720, but puzzling to admirers of the mathematical precision of Newton's *Principia* (1687), as was his inclusion of the causes of fermentation among the problems of natural philosophy along with those of gravity, electricity, and magnetism. As will be seen later in this book, Newton's interest in fermentation was related to his fascination with alchemy,[4] an aspect of his career that has been more fully appreciated by historians of science relatively recently.[5] Moreover, to some twentieth-century biochemists, Newton's statement that "in Fermentation Particles of Bodies . . . are put into new Motions" may have carried a foretaste of the enzyme-catalyzed "activation" of a chemical process.

Human knowledge of the phenomena of fermentation is at least as old as agriculture. The conversion of the juice of crushed sweet grapes (must) into wine, with effervescence, was known to ancient Greeks before the days of Homer, and Greek colonists introduced viticulture into southern Gaul in about 600 B.C. A similar effervescence had been observed in ancient Mesopotamia in the action of yeast (leaven, Gr. ζύμη, zyme) on a cereal dough, or the manufacture of beer by the action of hops on moist cereals (barley, wheat).[6] As

---

[3] *Ibid.*, pp. 400–401.

[4] Read, J. (1939). *Prelude to Chemistry*, pp. 307–308. 2nd ed. London: Bell; Forbes, R. J. (1949). "Was Newton an alchemist?" *Chymia* 2, 27–36.

[5] Dobbs, B. J. T. (1975). *The Foundations of Newton's Alchemy or "The Hunting of the Greene Lyon."* Cambridge University Press (see review by K. Figala, *History of Science* 15 (1977), 102–137); Westfall, R. S. (1980). *Never at Rest. A Biography of Isaac Newton.* Cambridge University Press; Henry. J. (1988). "Newton, matter, and magic" in: *Let Newton Be!*, J. Fauvel et al. (eds), pp. 127–145. Oxford University Press; Dobbs, B. J. T. (1991). *The Janus Face of Genius. The Role of Alchemy in Newton's Thought.* Cambridge University Press.

[6] Forbes, R. J. (1954). "Chemical, culinary, and cosmetic arts" in: *A History of Technology*, C. Singer et al. (eds.), Vol. 1, pp, 238–298. Oxford: Clarendon Press; Levey, M. (1959). *Chemistry and Chemical Technology in Ancient Mesopotamia.* Amsterdam: Elsevier; McGovern, P. E. (2003). *Ancient Wine.* Princeton University Press.

applied to such natural or artificial processes, the Greek term was *zymosis*, and the craft was called *zymotechnia*. The English (or French) term *fermentation* was derived from the Latin *fervere* (to boil) or *fervimentatio* (to heat); the German word is *Gärung* (or *Gährung*). The active agent in the process is termed *ferment*, which is also used in German (*Ferment*). The ancient artisans observed that the vinous fermentation was accompanied by the formation, in the fermenting liquid, of a deposit which took the form sometimes of a sediment, sometimes of a scum on the surface. If wine was allowed to stand for a time it turned sour, to yield vinegar, the strongest acid known to antiquity. This souring of wine was considered to be comparable to the curdling of milk by rennet in the manufacture of cheese. These non-effervescent processes were included among the fermentations, as was the decay of plant and animal material (putrefaction, Gr. σηψις, sepsis) with the release of noxious odors. The natural process of digestion (concoction, Gr. πεψις, pepsis) was long considered to be analogous to the artificial fermentations. The violent effervescence associated with the transformation of grape must into wine or of a cereal dough into bread provided analogies from everyday experience in efforts to explain other natural transformations of matter (for example, the germination of seeds, the generation of metals) by assuming the similar action of ferments. In Newton's time, chemically inclined physicians and natural philosophers considered fermentation to be one of the most important chemical reactions. Thus, the occurrence of an effervescence in the reaction of acids with bases was taken by some chemists as evidence of a fermentation process.

The production methods in the manufacture of wine or beer changed little from Neolithic times until the advent of the Industrial Revolution about 1800. Thus, the crushing of grapes advanced from the use of feet to mechanical means, thermometers were used to control the temperature of fermentation and aging, and stainless-steel tanks came to be used for storage. The most significant contribution to the development of the production methods of wine and beer was Louis Pasteur's introduction of pure strains of yeast, which will be discussed later in this book. This achievement transformed the making of wine and beer from a speculative enterprise, whose outcome was uncertain, into a science-based activity.

In this book, I offer a sketch of the usage in the Mediterranean world and western Europe of the terms fermentation and ferment (or their Greek, Latin, Arabic, or German equivalent) in alchemical efforts and in subsequent controversies about the nature of alcoholic fermentation.

The first chapter deals with the transmission of Aristotle's *Meteorologica* and his theory of change in matter, the must of vinous fermentation or ripening of fruit serving as examples. In about 300 A.D. the theory became part of an emerging alchemy, in which the recipes of Egyptian metal workers, dyers, and other craftsmen acquired an overlay of mysticism derived from Neoplatonism, Hermeticism, and astrology, and whose principal objective had become the artificial change of metals such as lead into gold. After the Muslim conquest, the Arabic translations (ca. 800–1000 A.D.) of this alchemical lore were then translated into Latin for scholars in western Europe. The development during 1300–1600 of the technical arts of mining and mineralogy promoted the invention of new furnaces and stills, and encouraged the effort to devise a method involving fermentation for the preparation of an "elixir" that would change lead into gold and also cure many human diseases. During the sixteenth century, Paracelsus agitated for the treatment of disease with chemical drugs, Sennert offered a corpuscular theory of matter, and Libavius published an important treatise on chemistry.

The second chapter begins with the work and thought of Van Helmont, who attached great importance to fermentations in the human body. Although given to mysticism, he made significant observations, notably the discovery of a "spiritus sylvestris" (which he named a "gas") during vinous fermentation. Among Van Helmont's several disciples were the Dutch physician dele Boë Sylvius, who worked on human digestion, and the English physician Willis, who interpreted fermentation as an "intestine motion" of corpuscles along the lines of the mechanical philosophy espoused by Descartes, Boyle and Newton. Becher and Stahl made a distinction between fermentation and putrefaction. Of special importance was the rediscovery in the eighteenth century of Van Helmont's "wild spirit" by Black, who named it "fixed air" and the work of Scheele on the isolation of organic acids from plant extracts.

Lavoisier's celebrated experiment on yeast fermentation opens the third chapter, which sketches the development of the problem during the nineteenth century. Beginning with the work of Dalton, Gay-Lussac, and Berzelius, there ensued a remarkable development of organic chemistry, with the concepts of valence, structure, and configuration leading at the end of the century to Fischer's synthesis of the sugars, and his demonstration of the stereochemical specificity of ferment action. In the interim a series of individual ferments (for example, pepsin, diastase) were identified, and chemical theories of fermentation were proposed by Berzelius and Liebig. These theories were opposed by Pasteur, whose microbiological studies led him to a vitalist view of yeast fermentation. The Liebig-Pasteur debate overshadowed Traube's prescient argument that ferments are chemical substances in living cells. A distinction was made between "organized ferments" and "unorganized ferments" and Kühne named the latter (e.g., "pepsin") "enzymes."

The fourth chapter begins with Buchner's 1897 report of the preparation, from brewer's yeast, of a cell-free aqueous extract ("zymase") which fermented glucose to ethanol and carbon dioxide. The experimental study of the action of zymase before World War I by Harden, Wróblewski, Ivanov, and others revealed that yeast fermentation is a multi-step chemical process involving several separate enzymes and the intermediate formation of phosphate derivatives of glucose, fructose, and glyceric acid. The initial pathways proposed by Wohl and Neuberg favored methyl glyoxal as a key intermediate. During the 1920s, research on enzymes was clouded by Willstätter's insistence that they were low-molecular-weight catalysts adsorbed on nonspecific colloids, and by his dismissal of Sumner's claim to have isolated an enzyme (urease) as a crystalline protein. The protein nature of enzymes was widely accepted only after Northrop's crystallization of pepsin in 1930. During the 1920s, two other prominent scientists (Warburg and Wieland) also contributed to the biochemical confusion in their debate about the nature of biological oxidation reactions. A resolution of the problem of yeast fermentation only came during the 1930s, with the replacement of methyl glyoxal by pyruvic acid as a key intermediate, and the demonstration that the pathway for the conversion of glucose to pyruvic acid is the same as that in anaerobic breakdown of glucose in mammalian muscle. The experimental evidence was largely provided by the research groups associated with Embden, Meyerhof, and Parnas, and the individual enzyme proteins in the so-called EMP twelve-enzyme pathway of yeast fermentation were isolated in crystalline form by the Warburg group.

CHAPTER ONE

ARISTOTLE TO PARACELSUS

The beliefs and practices of physicians and other ancient craftsmen (wine-makers, brewers, farmers, dyers, metal workers, miners) were reflected in the theories of Greek philosophers about the fundamental units of matter. Thales (fl. 585 B.C.) assumed that the basic principle is water, Anaximander (fl. 546 B.C.) that it is *apeiron* (the unbounded), and Anaximenes (fl. 546 B.C.) that it is *pneuma* (breath). About a hundred years later, Empedocles (492–432 B.C.) proposed that all things have four "roots"—fire and air (which rose upward), water and earth (which fell downward)—associated with the "active qualities" hot and cold, and the "passive qualities" wet and dry. Also, Leucippus (fl. 430 B.C.) and Democritus (fl. 420 B.C.) defined atoms as hard, indivisible particles of variable size moving in empty space.[1]

In his *Timaeus*, Plato (427–347 B.C.) rejected the ideas of Empedocles and the materialist atomists, and used Pythagorean numerology and geometry to argue that the fundamental entities were not units of matter but ideal Forms created by God. Plato assigned the four-sided pyramid to fire, the eight-sided octahedron to air, the twenty-sided icosahedron to water, and the six-sided cube to earth.[2] The only reference to fermentation is in a section dealing with the physiology of taste:

> There are other particles which, previously refined by putrefaction, enter into the narrow veins, and being duly proportioned to the particles of earth and air which are there, set them whirling about one another and enter into one another, and so form hollows surrounding the particles that enter. These watery vessels of air—for a film of moisture, sometimes earthy, sometimes pure, is spread around the air—are hollow spheres of water, and those of them which are pure are

[1] Freeman, K. (1949). *The Pre-Socratic Philosophers*. Oxford University Press.

[2] Plato (1961). *The Collected Dialogues*. E. Hamilton and H. Caims (eds.), pp. 1181–1182. New York: Pantheon Books.

transparent and are called bubbles, while those composed of the earthy liquid, which are in a state of general agitation and effervescence are said to boil or ferment.[3]

Plato's concept of a "world soul" and idea of the generation of metals from ice in the earth were highly regarded by the founders of alchemy, and his name was attached to some of their writings.[4]

Aristotle (384–322 B.C.) accepted the four elements but added a version of Plato's Forms as a fifth "essence" that gives a material thing its "soul," makes it more "complete," and brings it closer to the celestial world, where objects move in perfect circles. He provided a definition of a homogeneous substance, and defined "element" as a "body into which other bodies may be analyzed, present in them potentially or in actuality (which of these is still disputable), and not itself divisible into bodies different in form" (*De Caelo*, 302). Of particular importance in relation to Aristotle's influence on later natural philosophers and alchemists was his theory of material change. In his *Meteorologica*, he defined concoction (*pepsis*) in terms of two active opposites, hot and cold, and two passive ones, moist and dry:

> Concoction is maturity, produced from the opposite, passive characteristics by a thing's own natural heat, these passive characteristics being the matter proper to the particular thing. For when a thing has been concocted it has become fully mature. And the maturing process is initiated by the thing's own heat, even though external aids may contribute to it: as for instance baths and the like may aid digestion, but it is initiated by the body's own heat. In some cases the end of the process is a thing's nature, in the sense of its form and essence. In others the end of concoctions is the realization of some latent form, as when moisture takes on a certain quality and quantity when cooked or boiled or rotted or otherwise heated; for then it is useful for something and we say it has been concocted. Examples are must, the pus that gathers in boils, and tears when they become rheum; and so on.[5]

In his *Historia Animalium*, there is the following passage:

> Rennet is a sort of milk, it is formed in the stomach of young animals while being suckled. Rennet is thus milk which contains fire,

[3] *Ibid.*, p. 1190.

[4] Waley Singer, D. (1946). "Alchemical texts bearing the name of Plato," *Ambix* 2, pp. 115–128.

[5] Aristotle (1952). *Meteorologica* (translated by H. D. P. Lee), p. 299. Cambridge, Mass.: Harvard University Press.

which comes from of the heat of the animal while the milk is undergoing concoction.[6]

Aristotle took the word *pepsis* from the Hippocratic writings about human digestion and the concept of opposites from Empedocles.[7] He generalized the concept of *pepsis* to include the ripening of fruit, the development of an embryo, or the spontaneous generation of living things in the earth. For Aristotle, *pepsis* is a perfecting, by a thing's own innate (vital) heat, but can be exerted by external heat, and he offered as examples what is concocted in must (unfermented grape juice) or the fate of food carried by the blood to the heart, liver, or spleen. He differentiated natural *pepsis* from those promoted by moderate and strong heat. In Aristotle's philosophy, natural concoction generated by the vital heat expresses the tendency of an object to function toward a specific end; in a living thing, this property (*psyche*, soul) is inherent in the organism as a whole, and arises from the integrated functions of its parts.

He also invoked the existence of a connate *pneuma* as the substrate of the process initiated by the vital heat, and as a formative agent in the living organism.[8] Later, Aristotle's idea of *psyche* became mingled with the idea of *pneuma* expounded by the Stoics as a principle of cohesion and activity in both living and nonliving matter.

A noted Aristotle scholar has described the fourth volume of the *Meteorologica* as "the oldest extant chemical treatise. No distinction is made between organic and inorganic processes."[9] The technical language of the *Meteorologica* expresses the experience of generations of craftsmen. It is surprising, therefore, to find in the introduction to an English translation the following: "That the *Meteorologica* is a little-read work is no doubt due to the intrinsic lack of interest of its contents. Aristotle is so far wrong in nearly all his conclusions that they can, it may with justice be said, have little more than a passing antiquarian interest."[10] Much may have turned out to be wrong, such

---

[6] Aristotle (1965). *Historia Animalium* (translated by A. L. Peck), p. 229. Cambridge, Mass.: Harvard University Press.

[7] Boylan, M. (1982). "The digestive and 'circulatory' systems in Aristotle's biology," *Journal of the History of Biology* 15, pp. 89–118.

[8] Peck, A. L. (1953). "The cognate *pneuma*," *Science Medicine and History*, E. A. Underwood (ed.). Vol. 1, pp. 111–121. Oxford University Press.

[9] Düring, I. (1966). *Aristoteles*, p. 382. Heidelberg: Carl Winter.

[10] Aristotle (1952) (note 11), pp. xxv–xxvi. For another cavalier treatment of *Meteorologica*, see F. Salmon (1960). *Aristotle's System of the Physical World*, p. 403. Ithaca, New York: Cornell University Press.

as the idea that the earth's heat initiates the generation of metals, but the idea provided a stimulus for the centuries-long search of an agent ("ferment," "elixir," or "philosopher's stone") for the transmutation of lead into gold. Recent important studies have revealed much of interest, apart from its influence on alchemists, in Aristotle's enlargement of the scope of concoction.[11]

The writings attributed to Aristotle were collected and edited in about 70 B.C. by Andronicus of Rhodes, the eleventh Scholarch of the Lyceum in Athens. One of the last members of this school was Alexander of Aphrodisias; in about 200 A.D. he prepared commentaries on several of Aristotle's works, including Book 4 of the *Meteorologica*.[12] Such commentaries were important means of the later transmission of Greek natural philosophy to Arabic alchemists, notably Jābir ibn Hayyām, who appears to have lived in Baghdad during the early ninth century.[13]

Before the conquest of Syria, Persia and Egypt by Islam, the city of Alexandria had become the leading Mediterranean center of learning (also industry and commerce), and a melting pot of the religious beliefs of the Egyptians with those of the many Greeks, Persians, and Jews who lived there. Alexandria was probably the birthplace of Western alchemy, for the first authentic alchemical work, by Zosimos, appears there in about 300 A.D. He provided a compendium of both the occult and practical aspects of alchemy, with descriptions of the equipment (ovens, distilling apparatus).[14] His text

---

[11] Furley, D. (1989). "The mechanics of *Meteorologica* IV: A prolegomenon to biology," *Cosmic Problems*, pp. 132–148. Cambridge University Press; Freudenthal, G. (1995). *Aristotle's Theory of Material Substance. Heat and Pneuma, Form and Soul.* Oxford: Clarendon Press; Lloyd, G. E. R. (1996). "The master cook" in: *Aristotelian Explorations*, pp. 83–103. Cambridge University Press; Viano, C. (1996). "Aristote et l'alchimie grecque. La transmutation et le modèle aristotélien entre théorie et pratique," *Revue d'Histoire des Sciences* 49, pp. 189–213; Newman, W. R. (2001). "Experimental corpuscular theory in Aristotelian alchemy" in: *Late Medieval and Early Modern Corpuscular Matter Theory*, C. Lüthy et al. (eds.), pp. 291–329 (307–317).

[12] Lewis, E. (1996). *Alexander of Aphrodisias. On Aristotle Meteorology 4.* London: Duckworth.

[13] Plessner, M. (1973). "Jabir ibn Hayyam," *Dictionary of Scientific Biography* 7, pp. 39–43. New York: Scribners; Kraus, P. (1986). *Jābir ibn Hayyām.* Paris: Les Belles Lettres; Wilson, C. A. (1988). "Jabirian numbers, Pythagorean numbers, and Plato's *Timaeus*," *Ambix* 35, pp. 1–13; Haq, S. N. (1994). *Names, Natures and Things. The Alchemist Jābir ibn Hayyām and his* Kitāb al-Ahjār *(Book of Stones)*. Dordrecht: Kluwer.

[14] Hammer-Jensen, I. (1921). *Die älteste Alchymie*, pp. 98–125. Copenhagen: Høst; Taylor, F. S. (1930). "A survey of Greek alchemy," *Journal of Hellenic Studies* 50, pp.

included lengthy quotations from earlier alchemists, notably Maria the Jewess, whose achievements included the invention of a water-bath, now known (in France) as a *bain-Marie.*[15] Zosimos also referred to earlier alchemical writings attributed to Democritus (the pre-Socratic atomist), but many of them must have been written later. It was thought that an Egyptian philosopher, Bolos of Mendes, who lived in Alexandria around 130 B.C. was this pseudo-Democritus, but this view has been questioned.[16]

In the writings of Zosimos, the Greek tradition derived from Plato and Aristotle had been transformed by Neoplatonism, with emphasis on the *pneuma* of the Stoics and the arcane mysticism of Hermes Trismegistus, the Kabbalah, and Zoroaster. This attempt to make an esoteric doctrine out of the many chemical recipes left by craftsmen for the working and dyeing of metals, glass, pottery, cement, and cloth represented the appearance of alchemy in the Mediterranean world.[17] Among these recipes were those for the imitation of purple dyes and of gold by tinting a baser metal such as copper or tin by red iron oxide. The early Greek alchemists considered the color of a metal (along with other properties) to reveal its inner spirit, and used these recipes in the search for an agent analogous to yeast which would "elevate" the color of a metal in the series black, white, yellow, red. For example:

> If you wish to tint into silver, add leaves of silver; if into gold, leaves of gold. For Democritus says: Project Water of Sulfur on common gold and you can give it a perfect tint of gold. A single liquid is recognized as acting on both metals. It is necessary, therefore, that the Sulfur Water play the part of a yeast, producing the like, whether silver or gold. In fact, just as yeast, although in small quantity, raises a great quantity of dough, so also a little quantity of gold or silver acts by aid of this reagent.[18]

109–139; Hopkins, A. J. (1938). "A study of the kerotakis process as given by Zosimos and later chemical writers," *Isis* 29, pp. 326–354; Lindsay, J. (1970). *The Origins of Alchemy in Graeco-Roman Egypt*, pp. 323–3 57. London: Frederick Muller.

[15] Patai, R. (1994). *The Jewish Alchemists*, pp. 60–91. Princeton University Press.

[16] Hammer-Jensen, I. (1921) (note 20), pp. 80–98; Fraser, P. M. (1972), *Ptolemaic Alexandria*, Vol. 1, pp. 440–444, Vol. 2, pp. 636–646. Oxford University Press; Hershbell, J. P. (1987). "Democritus and the beginnings of Greek alchemy," *Ambix* 34, pp. 5–20.

[17] Stillman, J. M. (1924). *The Story of Early Chemistry*. London: Constable.

[18] Quoted from Hopkins, A. J. (1934). *Alchemy Child of Greek Philosophy*, p. 76. New York: Columbia University Press.

The wording of this recipe offers a foretaste of the language of alchemy, with its obfuscation and symbolic hints. It marks the beginning of a centuries-long effort to make an esoteric doctrine or philosophy out of the labors of craftsmen, with a rule of secrecy.

The kitchens of antiquity provided the basis for the ovens and other equipment of the alchemical laboratory.[19] The Alexandrian chemists made an important advance in building the first distilling apparatus with condensers.[20] For the alchemists this permitted the separation of the "pneuma" (spirit) from the contaminating "somata" (body). The improvement in the methods of distillation led to the production from wine of ethyl alcohol (*aqua ardens, aqua vitae*) during the course of the search for a "Quintessence", an alchemical addition to Aristotle's four elements.[21] The chemical and medicinal knowledge, as well as the experimental methods adopted by the Islamic alchemists, were largely derived from the work of Alexandrian craftsmen and the herbalist Dioscorides.[22] The Islamic scholars also acquired the sayings of Hermes Trismegistos, which included what was later translated into Latin as "fermentum" or English as "ferment":

> The Gold is their 'Divine Water'; and the 'Divine Water' is the 'Ferment' of the 'Bodies'; and the 'Bodies' are their 'Earth'. The 'Ferment' of the 'Divine Water,' which is the 'Ferment' of the 'Bodies,' is the Ash, and it is the 'Ferment of Ferments.'[23]

I offer this quotation as an example of hermetic language. What later appeared in Latin as "fermentum" was 'al-Iksir' in Arabic, and became "elixir."

The period 750–900 was the time of the Syrian translators who turned Aristotle's writings into Arabic. Jabir ibn Hayyam (fl. ca. 900) accepted Aristotle's doctrine of "opposites" and claimed that the transmutation of "elements" was possible only when a proper balance

---

[19] Halleux, R. (1981). *Les Alchimistes Grecs.* Vol. 1. Paris: Les Belles Lettres.

[20] Forbes, R. J. (1948). *A Short History of the Art of Distillation.* Leiden: Brill.

[21] Lu Gwei-Djen, J. Needham and D. Needham (1972). "The coming of ardent water," *Ambix* 19, pp. 69–112; Taylor, F. S. (1953). "The idea of the quintessence" in: *Science, Medicine and History*, E. A. Underwood (ed.), vol. 1, pp. 247–265. London: Oxford University Press.

[22] Lippmann, E. von (1923). "Die chemischen Kenntnisse von Dioscrorides" in: *Abhandlungen und Vorträge zur Geschichte der Naturwissenschaften.* Vol. 1, pp. 47–73. Berlin: Springer.

[23] Stapleton, H. E., G. L. Lewis, and F. S. Taylor (1949). "The sayings of Hermes quoted in the *Mā Al-Waraqī* of Ibn Umail," *Ambix* 3, pp. 69–90 (72).

of the "qualities" had been achieved. He considered metals to be composed of "sulfur" and "mercury," and thought that gold, the most "perfect" metal, is a combination of their completely "pure" forms. He seems to have known of ammonia and sal ammoniac. Jabir was followed by Abu Bakr Muhammad ibn Zkariyya al-Razi (ca. 854–925), a noted physician who believed that transmutation was possible, but rejected the idea of "balance." Al-Razi considered the purpose of alchemy to find "elixirs" which would convert base metals into gold and convert glass into precious stones. His chief contribution was the listing of the many known chemical substances.[24]

The most important Arabic successor of Jabir in the study of Aristotle's *Metereologica* was Abdallah ibn Sina (980–1037). In a book with a section on minerology, chemistry, and geology, ibn Sina argued that the gold produced by tincture was only an imitation. When this section was translated into Latin under the title *De Congelatonie et Conglutinatio Lapidum*, it was appended to a translation of *Metereologica* 4, leading many to think it to be by Aristotle.[25]

The Muslim alchemists distilled practically every available mineral and animal matter; the typical result was that a low fire raised a vaporous "spirit" which, upon condensation, became a liquid (a "water"), a stronger fire raised an oily inflammable fluid, and left in the alembic a "fixed" dry residue termed an "earth" or a "salt."

During the twelfth century, several Western scholars, notably Robert of Chester (fl. ca. 1140), Adelard of Bath (fl. ca. 1130), Vincent of Beauvais (ca. 1190–1264), and especially Gerard of Cremona (ca. 1114–1187), translated Arabic alchemical works into Latin. These writings were attributed to Jabir ibn Hayyam (latinized as Geber), al-Razi (Rhazes), ibn Sina (Avicenna), and some of these Latin versions were not translations from the Arabic but, as in case of pseudo-Geber, written about 1300 by European authors. In what follows I cite some writings which indicate that the alchemical concept of the term "ferment" was carried forward from Zosimos to the Western

[24] Heym, G. (1938). "Al-Razi and alchemy," *Ambix* 1, pp. 184–191; Partington, J. R. (1938). "The chemistry of Al-Razi," *ibid.*, 1, pp. 192–196; Holmyard, E. J. (1957). *Alchemy*, pp. 85–89. Harmondsworth: Penguin.

[25] Ruska, J. (1934). "Die Alchemie von Avicenna," *Isis* 21, pp. 14–51; Wickens, G. M. (ed.) (1952). *Avicenna: Scientist and Philosopher*. London: Luzac; Newman, W. R. (1989). "Technology and alchemical debate in late Middle Ages," *Isis* 80, pp. 423–445.

alchemists of the fourteenth century. For example, in a work translated by Gerard of Cremona, and attributed to al-Razi, there appears the phrase "et est fermentum duorum exir rubei et albi"[26] (exir = elixir). In another translation, it is stated that "[Gold] is the most effective Elixir and most comparable to the yeast which leavens the dough."[27] In the writings of pseudo-Geber, "the silver ferment 'ad azymum' is prepared by dissolving silver in its solvent . . . The gold ferment is prepared by dissolving gold in its solvent."[28]

During the period 1300–1600 there appeared several works whose authorship was attributed to famous philosophers or alchemists. Among them was the *Summa perfectionis* of pseudo-Geber.[29] Another was the *Turba philosophorum*, a collection of writings said to be by pre-Socratic philosophers who had little to say about ferments,[30] but a treatise by pseudo-Thomas Aquinas said much about "fermentum album et rubeum."[31] In his treatise on stones and metals, Albertus Magnus (ca. 1200–1280) wrote that their implanted power is indirect because it "goes through the intermediary of the elements and the fermentation,"[32] and in his *Mirror of Alchemy* Roger Bacon (ca. 1219–1292) wrote: "As in the making of bread, a little leaven nourisheth and fermenteth a great deal of Paste: so will the Philosopher that our stone bee fermented, that it may bee ferment to the multiplication of the stone."[33] During the thirteenth and fourteenth centuries learned members of religious orders became so interested in alchemy that Pope John XXII (reigning in Avignon) forbade such studies.[34]

---

[26] Steele, R. (1929). "Practical chemistry in the twelfth century. Rasis de aluminibus et salibus," *Isis* 12, pp. 10–46 (29).

[27] Ruska, J. (1935). *Das Buch der Alaune und Salze*, p. 33. Berlin: Verlag Chemie.

[28] Damstaedter, E. (1922). *Die Alchemie des Geber*, p. 12.2. Berlin: Springer.

[29] Newman, W. R. (1985). "New light on the identity of 'Geber'," *Sudhoffs Archiv* 69, pp. 76–90; (1991). *The* Summa perfectionis *of pseudo-Geber*. Leiden: Brill.

[30] Plessner, M. (1954). "The place of the *Turba philosophorum* in the development of alchemy," *Isis* 45, pp. 331–338.

[31] Goltz, D., J. Telle, and H. J. Vermeer (1977). *Der Alchemistische Traktat 'von der Multiplikation'*, p. 78. Wiesbaden: Franz Steiner.

[32] Riddle, J. M. and J. A. Mulhallond (1980). "Albert on stones and minerals" in: *Albertus Magnus and the Sciences*, J. A. Weisheipl (ed.), pp. 203–234 (208). Toronto: Pontifical Institute of Mediaeval Studies.

[33] Bacon, R. (1992). *The Mirror of Alchemy*, S. J. Linden (ed.), p. 22. New York: Garland.

[34] Partington, J. R. (1937), "Albertus Magnus on alchemy," *Ambix* 1, pp. 3–20; "Trithemus and alchemy," *Ambix* 2, pp. 53–59.

Of special interest is a fourteenth-century work *Pretiosa Margarita Novella* by Petrus Bonus of Ferrara (fl. ca. 1330), and which appeared in print in 1546. It has been translated into English, with the title *The New Pearl of Great Price*, from which I offer the following lengthy quotation:

> Of the ferment, which is the great secret of our Art, and without which it cannot attain its goal, the Sages speak only in the very obscurest terms. They seem to use the word in two senses, meaning either the elements of the Stone itself, or that which perfects and completes the Stone. In the first sense our Stone is the leaven of all other metals, and changes them into its own nature—a small piece of leaven leavening a whole lump. As leaven, though of the Same nature as dough, cannot raise it, until, from being dough, it has received a new quality which it did not possess before, so our Stone cannot change metals, until it is changed itself, and has added to it a certain virtue which it did not possess before. It cannot change, or colour, unless it have first itself been changed and coloured, as we learn from *Turba Philosophorum*. Ordinary leaven receives its fermenting power through the digestive virtue of gentle and hidden heat; and so our Stone is rendered capable of fermenting, converting, and altering metals by means of certain digestive heat, which brings out its potential and latent properties, seeing that without heat, as Theophilus tells us, neither digestion, operation, nor motion is possible . . . More difficult is the second sense of the ferment, which is the truly philosophical ferment, and wherein is the whole difficulty of our Art. For in this second sense it signifies that which perfects our Stone. The word ferment is derived from a root which denotes seething or bubbling, because it makes the dough rise and swell, and has a hidden dominant quality which prevails to change the dough into its own nature, rectifying and reducing it to a better and nobler state. It is composed of divers hidden virtues inherent in one substance. In the Same way, that ferment which is mixed with our quicksilver makes it swell and rise, and prevails to assimilate it to its own nature, thus exalting it into a nobler condition.[35]

Such accounts were preludes to the appearance of manuals for the transmutation of base metals in gold or silver. George Ripley (ca. 1415–1490) composed, in his *Compound of Alchemy*, a set of metrical stanzas describing the twelve "gates" to be passed: calcination, solution,

[35] Petrus Bonus (1894). *The New Pearl of Great Price* (translated by A. E. Waite), pp. 252–256. London: James Elliott. See Crisciani, C. (1973). "The conception of alchemy as expressed in the *Pretiosa Margarita Novella* of Petrus Bonus of Ferrara," *Ambix* 20, pp. 165–181.

separation, conjunction, putrefaction, congelation, cibation, sublimation, fermentation, exaltation, multiplication, projection. For example,

> For lyke as flower of Whete made into Past,
> Requireth Ferment whych Leven we call
> of Bred that yt may have the kyndly tast,
> and become Fode to Man and Woman most cordyall;
> Right so thy Medcyn Ferment thou shall
> That yt my tast wyth the Ferment pure,
> And all assays evermore endure.[36]

Other alchemists offered different instructions, for example: calcination, congelation, fixation, dissolution, digestion, distillation, sublimation, separation, inceration, fermentation, multiplication, projection.[37] Some of these terms are in use today to denote operations in a chemical laboratory. The terms fermentation, exaltation, and multiplication only have Hermetic significance, and the others refer to metallurgical operations.[38] The honest goldsmiths and metal workers declined to dabble with transmutation, but medieval Europe was beset with wars and revolts which ruined the economy and emperors, kings, and noblemen (especially in the German states) sought alchemists who promised to produce gold, and who went to another court when their failure (or fraud) was evident.[39]

Early in the sixteenth century there appeared Theophrastus Bombastus of Hohenheim (ca. 1493–1541), the self-styled Philippus Aureolus, Theophrastus Paracelsus, one of the most controversial figures in the history of chemistry.[40] His chief contribution was to reject the humoralism of Galen, and to argue that "It is not as they say, that alchemy makes gold, makes silver; here the project is to make *arcana* and to direct them against the diseases."[41] This advocacy of drinkable med-

---

[36] Ripley. G. (1652). "The Compound of Alchymie" in: *Theatrum Chemicum Britannicum*, E. Ashmole (ed.), pp. 107–187 (175). London: Nath. Brooke.

[37] Pernety, A. J. (1758). *Dictionnaire Mytho-Hermétique*, p. 99. Paris: Bauche. See also Ruland, M. (1964). *A Lexicon of Alchemy* (translated by A. E. Waite). London: Watkins.

[38] Karpenko, V. (1992). "The chemistry and metallurgy of transmutation," *Ambix* 39, pp. 47–62.

[39] Obrist, B. (1986). "Die Alchemie in der mittelalterlichen Gesellschaft" in: *Die Alchemie in europäischen Kultur und Wissenschaftsgeschichte*, C. Meinel (ed.), pp. 33–59. See also Lopez, R. S. (1953). "Hard times and investment in culture" in: *The Renaissance*, pp. 29–54. New York: Harper & Row.

[40] Pagel, W. (1958). *Paracelsus*. Basel: Karger.

[41] Quotation in Weeks, A. (1997). *Paracelsus. Speculative Theory and the Crisis of the Early Reformation*, p. 153. Albany: State University of New York Press.

icines containing metals (gold, antimony, mercury) was anticipated by John of Rupescissa during the fourteenth century.[42] In his manifold writings Paracelsus offered speculations based on three spiritual principles—"salt," "sulfur," and "mercury"—derived from Hermetic, Hellenistic, and Cabalistic thought, and the idea that the chemical transformations in the human body are effected by a spiritual entity he named the *Archaeus*.[43] He described, in cryptic language, the preparation of the philosopher's stone, whose powers are extolled as follows:

> The explanation of the power which the stone to drive out so many strange and wonderful diseases is not to be found in its complexion nor its specific form, nor indeed in its own proper character or other attribute, but arises from the subtle practice which is brought about by the preparations, reverberations, sublimations, digestions, separations, distillations, and manifold reductions and resolutions. All these operations constitute in the stone a power and acidity which it did not possess initially, but which were afterwards bestowed.[44]

About twenty years after Paracelsus died, there emerged a group of chemically-minded physicians, later named the "Paracelsians."[45] Some of them were also called "iatrochemists."[46] They compared the course of human diseases to the growth of minerals, and attributed volcanic eruptions to a fermentation process caused by the internal heat of the earth. Although much of the mysticism in Paracelsus's writings was abandoned, the Paracelsians retained some of his speculative ideas. The most important of his disciples was the Flemish physician Joan Baptista van Helmont (1579–1644). Others were Daniel Sennert (1572–1637), Alexander von Suchten (ca. 1520–1590), Joseph Du Chesne (also named Quercetanus, ca. 1544–1609), and Jean

---

[42] Multhauf, R. P. (1954). "John of Rupescissa and the origin of medical chemistry," *Isis* 45, pp. 359–367.

[43] Pagel, W. (1961). "The prime matter of Paracelsus," *Ambix* 9, pp. 117–135. See also Weeks, A. (1997) (note 47).

[44] Quoted from Sherlock, T. P. (1948). "The chemical work of Paracelsus," *Ambix* 3, pp. 33–63 (58).

[45] Multhauf, R. P. (1948). "Medical chemistry and 'the Paracelsians'," *Bull. Hist. Med.* 28, pp. 101–126; Webster, C. (2002). "Paracelsus, Paracelsism, and the secularization of the worldview," *Science in Context* 15, pp. 9–27.

[46] Scheider, W. (1972). "Chemistry and iatrochemistry" in: *Science, Medicine, and Society in the Renaissance*, A. G. Debus (ed.), pp. 141–150. New York: Science History Publications.

Beguin (ca. 1550–1620). Sennert advocated a corpuscular theory of matter, and offered experimental evidence for the view that natural and artificial fermentations involve separations of bodies into their smallest parts, and the reunion of these parts. In 1619, he wrote:

> Thus in milk there are butter, curds, and whey. And what else are digestions and coctions—not only those that are carried out by art, but also those that are brought about by nature in the bodies of plants and animals—than first a *diakrisis* and resolution of the bodies to be mixed into their minimal parts, and again a *synkrisis* and concretion according to the proper nature and use of each thing.[47]

Suchten conducted a quantitative experiment on the transmutation of copper into gold, and only got out as much gold as he had put in.[48] In his writings, Du Chesne defined fermentation as "a mixing of kindly matter for multiplication, or kindly seasoning, or leavening." He described many chemical medicines, and stated that

> there is also found in Saltpeter, a certaine Mercurial of ayerie nature, and which notwithstanding cannot take fyre, but is rather contrary thereunto . . . the which sowernesse is the general cause of *Fermentation*, and coagulation of natural things.[49]

Beguin established a pharmacy school in Paris, and his *Tyrocinium Chymicum* went through several editions.[50] He defined "digestion" as "an operation in which things are cooked by means of a digestive fire, just as in natural digestion meat is cooked in the stomach"[51] and "fermentation" as

> An exaltation of a thing in its substance; by means of digestion the active heat surpasses and changes the nature of what is passive . . .

---

[47] Quoted from W. R. Newman (2001). "Corpuscular anatomy and the tradition of Aristotle's *Meteorology*, with special reference to Daniel Sennert," *International Studies in the Philosophy of Science* 15, pp. 145–153 (151). See also Sennert. D., N. Culpeper, and A. Cole (1662). *Chymistry Made Easie and Useful* etc., pp. 155–157. London: Peter Cole.

[48] Haberling, W. (1929). "Alexander von Suchten," *Zeitschrift des Westpreussischen Geschichtsvereins* 69, pp. 177–228; Hubicki, W. (1960). "Alexander von Suchten," *Gesnerus* 44, pp. 54–63.

[49] Quercetanus, J. (1605). *The Practise of Chemicall, and Hermeticall Physicke.* Part 2, chap. 2. London: Creede.

[50] Patterson, T. S. (1937). "Jean Beguin and his *Tyrocinium Chymicum*," *Ambix* 2, pp. 243–298.

[51] Beguin, J. (1624). *Les Elements de Chymie*, p. 62. 3rd ed. Geneva: Jean Celerier.

> Liquid things which have inner warmth simply ferment themselves, as the juices of pears, apples or the must. But those which are cold require the addition of an external thing which can initiate effervescence and fermentation.[52]

Two other alchemists who were prominent around 1600 were Michael Sendivogius[53] and Basil Valentine. The latter does not appear to have existed, but his voluminous writings on practical chemistry (especially on antimony and its compounds) were "edited" by Johann Thölde.[54]

Among the antagonists of Paracelsus and the Paracelsians were the physician Thomas Erastus[55] (1523–1583), and Andreas Libavius (ca. 1560–1616), the author of one of the first textbooks of chemistry (*Alchemia*, 1597). Libavius adhered to the theory of the transmutation of metals. He defined fermentation as "the exaltation of a material in its essential part by means of a ferment which, by virtue of its spiritual nature, penetrates the entire mass, and converts it into its own nature . . . The ferment works chiefly by virtue of its inner heat."[56] He also wrote that "we are coagulated by God like cheese."[57]

At the beginning of the seventeenth century, the widely accepted definitions of fermentation and ferment were provided in the *Lexicon Alchemiae* by Martin Ruland (1532–1602), a disciple of Paracelsus:

> FERMENTATIO—The exaltation of a Matter into its essential part by means of a ferment which penetrates the entire mass, and operates therein in a peculiar manner, acting immediately on the spiritual

[52] *Ibid.*, p. 69.

[53] Holmyard, E. J. (1957). *Alchemy*, pp. 226–231. Harmondsworth: Penguin; Porto, P. A. (2001). "Michael Sendivogius on nitre and the preparation of the philosophers' stone," *Ambix* 48, pp. 1–16.

[54] Partington, J. R. (1961). *A History of Chemistry*. Vol. 2, pp. 185–203. London: Macmillan; Principe, L. (1987). "'Chemical translation' and the role of impurities in alchemy: Examples from Basil Valentine's *Triumphwagen*," *Ambix* 34, pp. 21–30; Priesner, C. (1986). "Johann Thölde und die Schriften des Basilius Valentinus" in: *Die Alchemie in der europäischen Kultur- und Wissenschaftsgeschichte*, C. Meinel (ed.), pp. 107–118. Wiesbaden: Harrassowitz; (1997). "Basilius Valentinus und Labortechnik um 1600," *Berichte zur Wissenschaftsgeschichte* 20, pp. 159–172.

[55] Karger, J. (1957), "Thomas Erastus (1524–1583), der unversöhliche Gegner des Theophrastus," *Gesnerus* 14, pp. 1–13.

[56] My translation of the modern German in *Die Alchemie des Andreas Libavius*, pp. 103–104. Weinheim: Verlag Chemie. See also Newman, W. R. (1999). "Alchemical symbolism and concealment: The chemical house of Libavius" in: *Architecture of Science*, P. Galison and E. Thompson (eds.), pp. 59–77. Cambridge, Mass.: MIT Press.

[57] Quoted from Newman, W. R. (2001) (note 17), p. 314.

> nature . . . Or:—Fermentation is the incorporation of a fermenting substance with a substance which is to be fermented. For even as a small modicum of ferment, or yeast, can leaven a large mass of flour, so does the chemical ferment assimilate itself to the thing that is to be fermented. Whatsoever be the nature of the ferment, of such is the fermented matter. By ferment the philosophers understand a true body and a true matter, which, united to its proper Mercury, convert it into the nature thereof.[58]

During the sixteenth century, small German books (*Bergbüchlein*, *Probierbüchlein*) became available to miners, metal workers, and assayers. The first systematic practical text, entitled *Pirotechnia*, by Vannoccio Biringuccio (1480–1537), appeared in 1540, and was followed by *De Re Metallica*, by Georgius Agricola (Georg Bauer, 1494–1555) in 1556. Later practical chemical books were those of Bernard Palissy (ca. 1510–1589) and Lazarus Ercker (ca. 1530–1594). These works were intended for craftsmen and artisans, and ridiculed alchemy. Thus, Biringuccio wrote of those who attempt to transmute base metals into gold

> . . . that those workers who so eagerly follow after and seek it proceed by just two pathways. One is that which takes its enlightenment from the words of wise philosophers, by means of which they think to attain it. This they call the just, holy, and good way, and they say that in this they are but imitators and assistants of Nature, indeed, manipulators and physicians of mineral bodies, purging them of superfluities and assisting them by augmenting their virtue and freeing them from their defects. In this way they sometimes proceed to corrupt these bodies in order to be able to separate the elements they contain as to reduce them, or sometimes they convert them into new substances by means of this art or by adding another spirit different from the first one. Thus they seek in this way to bring these materials to a certain point of corruption or to a separation of elements, or to remove or add spirits to things, or to make coarse materials subtle, and sometimes to make subtle things coarse. Therefore, as you can understand, these persons, with bridle broken, run a circular track night and day, without ever having rest; and surely I do not know whether anyone has ever arrived at the desired goal . . . [I]t can be said in conclusion that this art is the origin and foundation of many other arts, wherefore it should be held in reverence and practiced. But he who prac-

[58] Ruland, M. [1612] (1964). *A Lexicon of Alchemy* (translated by A. E. Waite), pp. 144. London: John M. Watkins. The book may have been compiled by Ruland the Elder's son Martin (1569–1611).

> tices it must be ignorant neither of cause nor of natural effects, and not too poor to support the expense. Neither should he do it from avarice, but only in order to enjoy the fine fruits of its effects and the knowledge of them, and that pleasing novelty which it shows to the experimenter in operation.[59]

Biringuccio was a contemporary of the physician Paracelsus, whose advocacy of chemical medicines and definition of chemistry as an analytical art evoked strong controversy. After his death in 1541, the manifold writings attributed to him attracted the attention of many European physicians, pharmacists, and natural philosophers, and Paracelsus gained the fame denied him during his lifetime. The most prominent of his posthumous disciples was Van Helmont, about whom more will be said in the next chapter. As noted above, Sennert advanced a corpuscular (atomic) theory of matter, thus coming before Pierre Gassendi (1592–1655) and Robert Boyle (1627–1691) in that regard. Some historians have considered Boyle to have been "a founder of scientific chemistry." Another noteworthy Paracelsian was Von Suchten, whose quantitative chemical work involved the use of an assayer's balance. His experiment on the recovery of gold, mentioned above, was evidence of the principle of the conservation of matter. For some historians, that principle was only established in 1789 by Antoine Lavoisier, in his account of an experiment on alcoholic fermentation, which will be discussed later in this book.

[59] Biringuccio, V. (1540). *Pirotechnia.* Venice. English translation by C. S. Smith and M. T. Gnudi (1943), pp. 336–337. Cambridge, Mass.: M.I.T. Press.

CHAPTER TWO

VAN HELMONT TO BLACK

The alchemical definition of fermentation occupied an important place in Van Helmont's natural philosophy.[1] According to the English translation of his *Ortus Medicinae*,

> Ferment is a formall created being, which is neither a substance nor an accident, but a neutrall thing framed from the beginning of the world in the places of its own Monarchie . . . I will not treat of Fables, and things that are not in being: but of Principles, and Causes, in order to their ends, actions and generations: I consider Ferments existing truly and in act, and individually by their kindes distinct.[2]

He adopted Paracelsus' *Archaeus* as the "workman of generation" to produce the "seed" by means of a ferment:

> Ferments are gifts, and Roots established by the Creator the Lord, for the finishing of Ages, sufficient, and durable, by continual increase, which of water, can stir up and make Seeds proper to themselves . . . the *Ferment* holds the Nature of a true Principle.[3] . . . the seed is a substance in which the Archaeus is already contained, a spiritual gas containing in it a ferment, the image of the thing, and moreover a dispositive knowledge of things to be done . . . one thing is not changed into another without a ferment and a seed . . . The ferment exhales an odour, which attracts the generating spirit of the Archaeus.[4]

Van Helmont dismissed the four principles of Empedocles and the sulfur-mercury-salt trio of Paracelsus, and offered his famous "wil-

---

[1] Partington, J. R. (1961), pp. 235–238; Pagel, W. (1982). *Joan Baptista Van Helmont*, pp. 79–87. Cambridge University Press.

[2] Quoted from Davis, A. B. (1973). *Circulation Physiology and Medical Chemistry in England 1650–1680*, p. 53. Lawrence, Kansas: Coronado Press.

[3] Quoted from Oldroyd, D. R. (1974). "Some Neo-Platonic and Stoic influences on mineralogy in the sixteenth and seventeenth centuries," *Ambix* 21, pp. 128–156 (141).

[4] Foster, M. (1924). *Lectures on the History of Physiology during the Sixteenth, Seventeenth, and Eighteenth Centuries*, pp. 135–140. Cambridge University Press; Partington, J. R. (1961) (note 60), p. 236.

low tree" experiment as proof of the theory that water is the ultimate principle in nature.[5]

Van Helmont was an able physician who believed, with Paracelsus, that the practice of medicine depended on the application of chemistry. He also believed, as a devout Christian and something of a mystic, that God had created nature, and His work was imitated by chemists, so that experimental chemical research gave knowledge of God.[6]

Among his personal researches was the study of the digestion of foodstuffs in the human body. In contrast to Galen's view that gastric digestion is a "concoction" effected by heat helped by acid, Van Helmont claimed that the effective specific agent was acid, helped by heat and a ferment provided by the spleen. He also proposed that several other "ferments" were active in other organs. Thus, in the fifth digestion, after passing through the duodenum and the liver, "the blood of the arteries is changed into the vital spirit of the Archaeus." These studies on digestion led Van Helmont to demonstrate that the sour taste of the acid in the stomach is relieved by a cream in the duodenum, which "doth straightway attain the savour of a salt," thus providing an indication that the duodenal cream was alkaline in nature. Some acid-alkali reactions were accompanied by an effervescence.[7]

Van Helmont also claimed to have discovered a volatile "wild spirit" (*spiritus sylvestris*) generated upon burning charcoal, and stated:

> I call this spirit, unknown hitherto, by the new name *Gas* which can neither be constrained by Vessels, nor reduced to a visible body, unless the seed being first extinguished. But Bodies do contain this spirit, and do sometimes wholly depart into such a Spirit not indeed, because it is actually in those very bodies (for truly it could not be detained, yea

---

[5] Hoff, H. E. (1964). "Nicholas of Cusa, Van Helmont, and Boyle: The first experiment of the Renaissance in quantitative biology and medicine," *J. Hist. Med.* 14, pp. 99–117; Webster, C. (1966). "Water is the ultimate principle of nature: The background to Boyle's Sceptical Chemist," *Ambix* 13, pp. 96–107.

[6] Heinecke, B. (1995). "The mysticism and science of Johann Baptist Van Helmont (1579–1644)," *Ambix* 42, pp. 63–78.

[7] Boas, M. (1956). "Acid and alkali in seventeenth century chemistry," *Archives Internatonales d'Histoire des Sciences* 34, pp. 13–22; Debus, A. G. (2001). *Chemistry and Medical Debate. Van Helmont to Boerhaave*, pp. 103–136. Canton, Mass.: Science History Publications.

> the whole composed body should flie away at once) but it is a Spirit grown together, coagulated after the manner of a body and is stirred up by an attained ferment, as in Wine, the juyce of unripe Grapes, bread, or water and honey etc.[8]

According to Van Helmont, other forms of this spirit, with a different odor or color, were released upon heating saltpeter, the action of aqua fortis on silver, or the burning of sulfur. He also claimed to have prepared a universal solvent (*Alkahest*) which cures every sort of illness.[9] A recent estimate of Van Helmont as an experimenter, however, is that "a chemist who claimed to have had visionary dreams could well be an avid experimentalist who believed that certainty in physical matters could best be acquired by means of exact weights and measures.[10]

The posthumous appearance in 1648 of Van Helmont's collected writings (edited by his son Franciscus Mercurius) under the title *Ortus Medicinae* was warmly welcomed, and was cited in numerous publications during the second half of the seventeenth century. An English translation was published in 1662. The popularity of the book matched that of the *Furni novi Philosophici* (published during 1646–1649) by the industrial chemist and entrepreneur Johann Rudolph Glauber (1604–1670), famous for his *sal mirabile*.[11] Although not a "philosophical" chemist like Van Helmont, he also sought a universal curative agent. In his *Libellus Ignium*, or *Book of Fires*, he wrote:

> One great Secret more, above others, which for brevity sake cannot be inferred there, is this. It is well known to all Chymists, that all Vegetables, as also Animals, by addition of common Water, may be brought to fermentation, and according to every subject, a *Spiritus Ardens*, of great use in Physick, may be distilled. But how to make such a subtile Spirit out of Metals, I never read in any Authour, nor heard

[8] Van Helmont, J. B. (1662). *Oriatrike* (translated by J. L. M.), p. 106. London: Lodowick Loyd. See Pagel, W. (1962). "The 'wild spirit' (gas) of John Baptist Van Helmont (1579–1644)," *Ambix* 10, pp. 1–13; (1982).

[9] Joly, B. (1996). "L'alkahest, dissolvant universel ou quand la théorie rend pensable une pratique impossible," *Revue d'Histoire des Sciences* 49, pp. 306–344.

[10] Newman, W. R. and Principe, L. M. (2002). *Alchemy Tried in the Fire*, pp. 56–91. University of Chicago Press.

[11] Pietsch, E. (1956). *Johann Rudolph Glauber*. Munich: Oldenbourg; Young, J. T. (1998). "Universal medicines: Johann Glauber in England" in: *Faith, Medical Alchemy and Natural Philosophy*, pp. 183–257; Smith, P. H. (2000). "Vital spirits, redemption, artisanship, and the new philosophy in early modern Europe" in: *Rethinking the Scientific Revolution*, M. J. Osler (ed.), pp. 119–135. Cambridge University Press.

> of. But I have found a way by which great matters may be done, which cannot be mentioned here, let others search after it as I had done, it is not good to cast Pearls before Swine. Yet that the desirous may know somewhat of an Artificial Metallick Fermentation; I say that out of certain Salts a water may be prepared, which when it is put upon a compact Metal, that the Same by a certain property in the Water, begins to swell up and ferment, like in the fermentation of Wine, Beer, or other vegetable Drinks; and after fermentation, by distilling in Balneo yields an exceeding subtile penetrating Spirit, strong, sweet, and volatile.[12]

It should be noted briefly here that earlier in the seventeenth century, fermentation was invoked without reference to "subtile Spirits." For example, in his uncompleted *Novum Organum*, published in 1620, Francis Bacon (1561–1626) wrote:

> Heat does not diffuse itself when heating other bodies by any communication of the original heat, but only by exciting the parts of the heated body to that motion which is the form of heat... So the leaven of bread, yeast, rennet and some poisons, excite and invite successive and continued motion in dough, beer, cheese, or of the human body; not so much from the power of the exciting, as the predisposition and yielding of the excited body.[13]

In 1633, the iatrochemist Angelo Sala (ca. 1576–1637) defined fermentation as "an internal motion of the particles of bodies brought about by internal heat in the presence of moisture, which groups them in new arrangements, partly by separation and partly by reduction into a new kind of more noble mixt."[14]

Shortly before his death, René Descartes (1596–1650), who derived his "clear and distinct" notions about material things solely in terms of shapes, sizes, and motions, and based on geometry and mechanics, wrote in his *De la Formation du Foetus*:

> ... the seed of animals, which, being very fluid and ordinarily produced by the coming together of sexes, seems to be only a mixture compounded of two liquors which, serving each as a ferment to the

---

[12] Glauber, J. R. (1689). *The Works* (translated by C. Parke). Part II, p. 218. London: Thomas Milboum.

[13] Bacon, F. (1620). *Novum Organum*, p. xlviii. London. See Farrington, B. (1953), "On misunderstanding the philosophy of Francis Bacon" in: *Science Medicine and History*, E. A. Underwood (ed.), vol. 1, pp. 439–454. Oxford University Press.

[14] Partington, J. R. (1961), p. 280. See Gantenbein, L. (1992). *Der Chemiater Angelus Sala 1576–1637*, p. 198. Zurich: Juris.

> other, are so heated that some of the particles, acquiring the Same agitation that fire has, move apart and press against others . . . For, as we see that old dough can make new dough rise, and that the foam that beer throws up suffices as a ferment for other beer, so it is easy to believe that the seminal liquids of the two sexes, being mingled, serve as ferments to each other.[15]

In his *Traité de l'Homme* (ca. 1631, published 1664), he considered digestion to be a fermentation.

Despite the increasing commitment to a "mechanical philosophy" and a "corpuscular" state of matter, the influence of Van Helmont's Neoplatonic thought was evident throughout the seventeenth century, with its multitude of fermental "spirits" in both material and nonmaterial form. For example, in 1632 Edward Jorden, an English physician, wrote as follows about the generation of metals in the earth:

> There is a Seminarie Spirit of all minerals in the bowels of the earth, which meeting with conuenient matter, and adiuuant causes, is not idle, but doth proceed to produce minerals, according to the nature of it, and the matter which it meets withall; which matter it workes upon like a ferment, and by its motion procures an actuall heate, as an instrument to further his work; which actuall heate is increased by the fermentation of the matter. The like we see in making of malt, where the graynes of Barley being moystened with water, the generative Spirit in them, is dilated, and put into action; and the superflutie of water, being removed, which might choake it, and the Barley laid up in heapes; the Seedes gather heat, which is increased by the contiguitie of many graines lyiing one upon another. In this worke matures intent is to produce more individuals, according to the nature of the Seede, and therefore it shootes forth in spyres: but the Artist abuses the intent of nature, and coverts it to his end, that is, to increase the spirits of his Malt. The like we finde in mineral substances, where this spirit or ferment is resident.[16]

The iatrochemist Franciscus de la Boë (1614–1672), who latinized his family name to Sylvius, followed Van Helmont in considering

[15] Hall, T. E. (1970). "Descartes' physiological method: position, principles, examples," *Journal of the History of Biology* 3, pp. 53–79 (68, 71).

[16] Jorden, E. (1631). *A Discourse of Naturall Bathes, and Minerall Waters*, pp. 57–58. London: Thomas Harper. See Debus, A. G. (1969). "Edward Jorden and the fermentation of metals. An iatrochemical study of terrestrial phenomena" in: *Towards a History of Geology*, C. J. Schneer (ed.), pp. 101–121; Oldroyd, D. R. (1974). "Some Neo-Platonic and Stoic influences on mineralogy in the sixteenth and seventeenth centuries," *Ambix* 21, pp. 128–156.

the physiology of digestion as largely a matter of fermentation, but he rejected the idea of an Archaeus. Also, he revised Van Helmont's sequence of fermentations; after 1650 anatomical studies by three of his pupils had shown the existence of the submaxillary gland (Thomas Wharton), the duct of the parotid gland (Nicolas Steno), and the pancreatic juice (Regnier de Graaf). Sylvius adopted Van Helmont's idea of an acid-alkali balance in digestion, and this idea was broadened by Otto Tachenius (1615–1680) to include all things: "*Fire* and *Water*, or *Acid* and *Alcaly* (call them which you will) is that *Balsam*, which given to Bodies for *Salt*... This is that innate *Calid*... which doth abound in things that grow because it is fermentable and expirable."[17]

Although Sylvius distinguished between the process of fermentation and the effervescence sometimes associated with it,[18] in 1688, the French physician Raymond de Vieussens (ca. 1635–1715) defined fermentation as "the adventitious and expansive movement of the heterogeneous parts and of insensible fermenting bodies excited without sensible cause, which, when it is vehement or of long duration, brings about an essential change or conspicuous alteration in the fermenting bodies," and his six types of fermentation are confused with effervescence.[19] To indicate the state of discourse about fermentation, I add that in 1651 John French (1616–1657) had stated: "Fermentation is when any thing is resolved into it self, and is rarified, and ripened, whether it be done by any ferment added to it or by digestion only."[20]

The most important follower of Van Helmont and Sylvius in the matter of fermentation was the Oxford physiologist Thomas Willis (1621–1675).[21] Like many others of his time and place, Willis had adopted the corpuscular theory of matter.[22] In his tract *De Fermentatione*

[17] Tachenius, O. (1690), *Clavis to the Ancient Hippocratical Physick or Medicine*, p. 79. London: Marshal.

[18] Foster (1924) (note 69), pp. 146–160; King, L. S. (1970). *The Road to Medical Enlightenment 1650–1695*, pp. 93–112, 134–136. London: Macdonald.

[19] Partington, J. R. (1961), pp. 290–291; Debus, A. G. (1991). *The French Paracelsians*, pp. 139–140. Cambridge University Press.

[20] French, J. (1651). *The Art of Distillation*, p. 10. London: Cates.

[21] Partington, J. R. (1961), pp. 305–310; Isler, H. (1964). *Thomas Willis*. Zurich; Frank, R. G. (1980). *Harvey and the Oxford Physiologists*, pp. 165–169. Berkeley: University of California Press; Debus, A. G. (2001), pp. 64–73.

[22] Hooykaas, R. (1949). "The experimental origin of chemical atomic and molecular theory before Boyle," *Chymia* 2, pp. 65–80; Clericuzio, A. (2000), *Elements, Principles and Corpuscles*. Dordrecht: Kluwer; Luthy, C. et al. (eds.) (2001) (note 17).

(1659), with an English translation in *Dr. Willis's Practice of Physick* (1684), he defined fermentation as

> an intestine motion of Particles, or the Principles of every Body, either tending to the perfection of the Same Body, or because of its change into another. For the Elementary Particles being stirred up into motion, either of their own accord or Nature, or occasionally, do wonderfully move themselves, and are moved; do lay hold of and obvolve one another: the subtil and more active, unfold themselves on every side, and endeavour to fly away; which not withstanding being intangled, by others more thick, are deteined (sic) in their flying away. Again, the more thick themselves, are very much brought together by the endeavour and Expansion of the more Subtil, and are attenuated, until each of them being brought to their height and exaltation, they either frame the due perfection in the subject, or compleat the Alterations and Mutations designed by Nature.[23]

After describing the role of "fermentative particles" in the generation of metals and minerals, the formation of meteors, and stating that seminal vessels "swell up with Fermentative Particles; that there is nothing more: here Spirit, Salt, and Sulphur being together compacted, and highly exalted, seem in the seed to be reduced as it were into a most noble Elixir." To this he added: "We are not only born and nourished by means of Ferments; but we also Dye: Every Disease acts its Tragedies by the strength of some Ferment" and concluded:

> Having thus far wandered in the spacious field of Nature, we have beheld all things full of Fermentations; not only in the distinct Provinces of Minerals, Vegetables and Animals . . . but also the whole sublunary World, seems as if one and the Same substance were planted, and very pregnant through the whole with Fermentative Particles; which in every Region and Corner of it, as little Emmits in a Mole-hill, are busied in perpetual motion and agitation, they fly about here and there; sometimes upwards, sometimes downward they are hurried, they variously meet one another, associate themselves, and again depart asunder; with a continued Vicissitude they enter into divers Marriages, and suffer Divorces, on which the beginnings, the death, and transformations of things depend.[24]

---

[23] Willis, T. (1681). *Of Fermentation*, pp. 9–11. London: Dring et al.
[24] *Ibid.*, p. 16.

Other English followers of Van Helmont were Thomas Browne (1605–1682) who, like Sylvius, equated fermentation with effervescence,[25] and Walter Charleton (1620–1701) who wrote about natural vinous fermentation as follows:

> As for the Principal *Agent*, or *Efficient Cause* of this operation I perswade my self, You will easily admit it to be other but the *Spirit* of the Wine it self. Which, according to the Mobility of its nature, seeking after liberty, restlessly moving every way in the mass of liquor, thereby dissolves that common tye of mixture, whereby all the Heterogeneous parts thereof were combined and blended together, and having gotten it self free, at length abandons them to the tendency of their gravity and other properties. Which they soon obeying . . . leave the liquor to the possession and government of its noblest principle, the spirit. As for this spirit, as it is the life of the Wine, so doubtless it is also the cause of its Purity and Vigour, in which the perfection of that life seems to consist.[26]

Charleton is better known for his later advocacy of atomism.[27]

In 1674 there appeared *Tractatus Quinque Medico-Physici*, by John Mayow (1641–1679),[28] in which fermentation was defined as an interaction of a highly fermentative "nitro-aërial spirit" and the inflammable "sulfur" of material substances: "Nitro-aërial spirit and sulphur are engaged in perpetual hostilities with each other, and indeed from their mutual struggle when they meet and from their diverse state when they succumb by turns all the changes of things seem to arise."[29] Mayow defined the place of nitric acid ("nitrous spirit") by stating "With regard then to the aërial part of nitrous spirit, we maintain that it is nothing else than the igneo-aërial particles which are quite necessary for the production of any flame. Wherefore, let me call the fiery particles, which occur also in air, nitro-aërial particles or nitro-aërial spirit."[30]

---

[25] Merton, S. (1966). "Old and new physiology in Sir Thomas Browne: Digestion and some other functions," *Isis* 57, pp. 249–259.

[26] Charlton. W. (1659). *Two Discourses. I. Concerning the different Wits of Men. II. Of the Mysterie of Vintners*, pp. 147–148. London: Willaim Whitwood.

[27] Gelbart, N. R. (1971). "The intellectual development of Walter Charlton," *Ambix* 18, pp. 149–168.

[28] Partington, J. R. (1956). "The life and work of John Mayow," *Isis* 47, pp. 218–230, 405–417; (1961), pp. 577–613; Böhm, W. (1963). "John Mayow and his contemporaries," *Ambix* 11, pp. 105–120.

[29] Mayow, J. (1907). *Medico-Physical Works*, p. 35 (Alembic Club Reprint No. 12). Edinburgh: Thin.

[30] *Ibid.*, pp. 13–14.

In experimental studies on the breathing of animals, Robert Hooke (1635–1703) had shown that they consumed a portion of common air before they died. Mayow extended these experiments and explained animal respiration as follows:

> For in the first place nitro-aërial spirit when mixed with the saline-sulphureous particles of the blood appears to excite in it vital fermentation. In fact, just as nitro-aërial particles when they slowly enter the pores of the earth encounter there saline-sulphureous particles, immature indeed. In an obscure fermentation on which, as has been shown elsewhere, the life of plants depends; so the same nitro-aërial particles when introduced more profusely into the blood by the action of the lungs, and mixed in their minutest parts with its saline-sulphureous particles, brought to a State of active vigour, produce a very marked fermentation such as is required for animal life . . . Indeed I attempted to show above that nearly all fermentations of natural things result from the motion of nitro-aërial particles; and in fact I have no doubt at all that the effervescence of the blood is due to the same cause; accordingly when respiration is arrested, the effervescence of the blood immediately ceases, and animal life is extinguished.[31]

Mayow also offered some comments about digestion: ". . . the vulgar opinion is that there is in the stomach a certain acid ferment . . . I conclude that the digestive liquid of the stomach is not very different kind from saliva . . . The food is concocted by the ferment of the stomach into chyme, which, when it has passed into the duodenum immediately meets the bile, by which, as by a new ferment mixed with it, it is further fermented and concocted."[32]

A less famous English chemical physician, and disciple of Van Helmont, was William Simpson (1636/7–1680), whose book *Zymologia physica* was published in 1675. For Simpson, "Fermentation it self, which whether in minerals or vegitables, is nothing else but an intestine *motion* of the essential constituents of *Acidum* and *Sulphur*."[33] "And as before we have shewed, the *Fermentation* of *minerals* to consist in the *collision* and *intestine wrestlings* of their *Acid* and *Sulphur*, as the causes of *hot Baths*, &c. So the *Fermentation* in *animals* is no less *observable* to proceed from the like inward *struglings* of their imbred *Acid*

[31] *Ibid.*, pp 101–102.
[32] *Ibid.*, pp. 264, 267.
[33] Poynter, F. N. L. (1953). "A seventeenth-century controversy. Robert Witty versus William Simpson" in: *Science Medicine and History*, E. A. Underwood (ed.). Vol, 2, pp. 72–81. Oxford University Press.

and *Sulphur* . . . in order to the keeping those *Ferments* at work, for the *nourishing* and *upholding* the *fabrick* of those *bodies*.[34] Simpson also invokes "*Acido-nitro-sulphurous*" interactions in the generation of thunder, lightning, and earthquakes.[35]

To the above disciples of Van Helmont must be added George Starkey (1628–1665), an American alchemist and physician. William Newman has recently provided strong evidence that the seventeenth-century alchemical writings under the pen name of Eirenaeus Philalethes had been composed by Starkey.[36] In seeking to prepare an Elixir, he adopted Van Helmont's precept that the "ferment" characteristic of a body is composed of tiny corpuscles which can penetrate the core of the larger ones, where it exerts its "fermentative" power.[37] John Webster (1611–1682), another Helmontian, wrote that in transmutation

> there is a radical Solution and Penetration of all the small parts or atoms of the metal to be changed, by the subtile permeability and the ingression of their so much purified and exalted Tincture, and whereby and thereby all things in it whatsoever that are of Heterogeneous nature, are separated and extruded, and the Homogeneous Particles joined together per minima as much as Nature can admit of and so must needs be of less bulk, and possesses less room or place, which is manifest in Gold.[38]

The most famous of the Helmontians was the noble and wealthy Robert Boyle (1627–1691), a writer of religious tracts and works in chemistry.[39] His most notable experimental achievement was the establishment of the quantitative pressure-volume relationship (Boyle's Law) by means of an air pump designed by Robert Hooke (1635–1703). In my opinion, Boyle's most significant chemical work was in analytical chemistry, especially in relation to color tests (*Experiments and Considerations Touching Colours*, 1664). Boyle adopted the corpuscular

[34] Simpson, W. (1675). *Zymologiaphysica*, p. 15. London: W. Cooper.
[35] *Ibid.*, p. 49.
[36] See Partington, J. R. (1961), p. 484.
[37] Newman, W. R. (1994). *Gehennical Fire. The Lives of George Starkey, an American Alchemist in the Scientific Revolution.* Cambridge, Mass.: Harvard University Press.
[38] Newman, W. R. and L. M. Principe (2002), *Alchemy Tried in the Fire. Starkey, Boyle, and the Fate of Helmontian Chemistry*, pp. 136–155. University of Chicago Press.
[39] Clericuzio, A. (1996). "Alchimie, philosophie corpusculaire et minéralogie dans la Metallographia de John Webster," *Archives Internationales d'Histoire des Sciences* 49, pp. 287–304 (301).

theory of matter advocated by Daniel Sennert (1572–1637), Pierre Gassendi (1592–1655) and Walter Charleton, and the mechanical philosophy of Francis Bacon and René Descartes. Boyle's most famous book *The Sceptical Chymist* (1661) deals largely with the problem of defining the "elements" which constitute chemical substances; there are also condemnations of "vulgar chymists" and Paracelsus, and frequent (mostly favorable) references to Van Helmont. Boyle's appeal in *The Sceptical Chymist* for the avoidance of secrecy in chemical publication was contradicted by his later use of various means to conceal the details of his alchemical work.[40]

As regards Boyle's views on fermentation, in his speculations about the generation of minerals, "he appears to restrict the operation of seeds for explaining the generation and growth of plants and animals and to account for the growth of minerals by something analogous to ferments."[41] He considered that "the production of spirit in fermentation is a consequence of the intestine motion, in which the corpuscles by jostling against one another are broken, variously ground, and subtilised until they are qualified to be raised by a gentle heat before the phlegm."[42] Boyle also stated that the physician who

> throughly understands the nature of Ferments and Fermentations, shall probably be much better able then (sic) he that ignores them, to give a fair account of diverse *phaenomena* of severall diseases (as well Feavers as others) which will perhaps be never throughly understood, without an insight into the doctrine of Fermentation.[43]

Boyle became an alchemist under the tutelage of George Starkey, and hoped that the "incalescent mercury" (an amalgam of mercury and gold formed with the liberation of heat) he prepared would provide a means for the transmutation of metals.[44]

---

[40] Maddison, R. E. W. (1969). *The Life and Works of the Honourable Robert Boyle F.R.S.* London: Taylor & Francis; Hunter, M. (ed.) (1994). *Robert Boyle Reconsidered.* Cambridge University Press.

[41] Principe, L. M. (1992). "Robert Boyle's alchemical secrecy: codes, ciphers, and concealments," *Ambix* 39, pp. 63–74.

[42] Anstey, P. R. (2002), "Boyle on seminal principles," *Studies in History and Philosophy of Biological and Biomedical Sciences* 33, pp. 597–630 (623).

[43] Partington, J. R. (1961), p. 545; Clericuzio, A. (1990). "A redefinition of Boyle's chemistry and corpuscular philosophy," *Annals of Science* 47, pp. 561–589.

[44] Boyle, R. (1999). *The Works of Robert Boyle*, M. Hunter and E. B. Davis (eds.), vol. 3, p. 321. London: Pickering and Chatto.

In 1691/1692, Isaac Newton composed a short paper *De Natura Acidorum*; the full text and English translation only appeared in 1961.[45] Its partial publication in 1710 by John Harris (ca. 1666–1719) in his *Lexicon Technicum* marked the public appearance of Newton's earlier commitment to a theory of the attraction of particles of matter and its role in fermentation:

> The particles of acids are coarser than those of water and therefore less volatile; but they are much finer than those of earth, and therefore much less fixed than they. They are endowed with a great attractive force and in this force their activity consists by which they disolve bodies and affect and stimulate the organs of the senses. They are of a middle nature between water and [terrestrial] bodies and they attract both . . . When they attracted and gathered together on all sides they raise, disjoin and shake the particles of bodies; and by their force of attraction by which they rush to the [particles of] bodes, they move the fluid and excite heat and shake asunder some particles to such a degree as to turn them into air and generate bubbles; and this is the reason of dissolution and violent fermentation . . . But the acid, suppressed in sulphureous bodies, by attracting the particles of other bodies (for example, earthy ones) more strongly than its own, cause a gentle and natural fermentation and promotes it even to the stage of putrefaction in the compound . . . Note that what is said by chemists, that everything is made from sulphur and mercury is true, because by sulphur they mean acid, and by mercury they mean earth.[46]

Newton's study of alchemy began during his student days at Trinity College, Cambridge, and in addition to collecting transcriptions, translations, and extracts of the alchemical literature, he prepared a chemical dictionary, in which he defined "fermentation" as

> the working of liquors, whereby they are further digested & seperated (sic) from their faeces &c. Tis impeded by cold. And Must immersed for 6 or 8 weeks in a cold well is soe satled in its constitution that it will not ferment of (sic) a long while after.[47]

[45] Principe, L. M. (1998). *The Aspiring Adept. Robert Boyle and his Alchemical Quest.* Princeton University Press; Newman, W. R. and L. M. Principe (2002). *Alchemy Tried in the Fire. Starkey, Boyle, and the Fate of Helmontian Chemistry.* University of Chicago Press.

[46] Newton, I. [1691/1692] (1961). "De Natura Acidorum" in: *The Correspondente of Isaac Newton*, H. W. Trumbull (ed.), Vol. 3 (1688–1694), pp. 205–214. Cambridge University Press.

[47] Newton, I. (1961) (note 111), pp. 209–210.

Newton performed many chemical experiments until about 1694.[48] He noted that

> . . . Nature seems delighted with transmutations. Water, which is a very fluid tasteless Salt, she changes by Heat into Vapour, which is a sort of Air, and by Cold into Ice, which is a hard, pellucid, brittle, fusible Stone, and this Stone returns into Water by Heat, and Vapour returns into Water by Cold. Earth by heat become Fire, and by Cold returns into Earth. Dense Bodies by Fermentation rarify into several sorts of Air, and this Air by Fermentation, and sometimes without it, returns into dense Bodies.[49]

Newton also asked whether "all bodies therefore abound with a very subtile, but active, potent electric spirit" and whether

> By the action of the same spirit some particles of bodies can the more strongly attract one another, others less strongly, and thence can arise varying congregations and separations of particles in fermentations and digestions, especially if the particles are agitated by slow heat.[50]

During 1650–1700, the intellectual life of Cambridge was filled with "spirits," and Newton's idea of the "spirit" may have derived an impetus from Mayow's nitro-aerial spirit.[51] In 1675 Newton wrote to Henry Oldenburg (ca. 1616–1677): "The whole frame of nature may be nothing but aether condensed by a fermentative principle."[52]

Although he believed in the possibility of transmutation of metals, Newton does not appear to have worked with that as a goal. Nevertheless, he was secretive and except for his assistant Humphrey Newton, almost no one was allowed in his laboratory.[53] His exten-

---

[48] Quoted from Dobbs, B. J. T. (1975) (note 5), p. 173.

[49] Boas, M. and A. R. Hall (1958). "Newton's chemical experiments," *Archives Internationales d'Histoire des Sciences* 11, pp. 113–152.

[50] Newton, I. (1730) (note 2), pp. 374–375. See McGuire, J. E. (1967). "Transmutation and immutability: Newton's doctrine of physical qualities," *Ambix* 14, pp. 69–95.

[51] Quoted from McGuire, J. E. (1968). "Force, active principles, and Newton's invisible realm," *Ambix* 15, pp. 154–208 (176–177).

[52] Schaeffer, S. (1987). "Godly men and mechanical philosophers: Souls and spirits in Restoration natural philosophy," *Science in Context* 1, pp. 5–85; Hall, A. R. (1998). "Isaac Newton and the aerial nitre," *Notes and Records Royal Society of London* 52, pp. 51–61.

[53] Newton I. (1959). *The Correspondence of Sir Isaac Newton.* Vol. 1, p. 414. Cambridge University Press.

sive chemical work with lead and mercury may have contributed to his illness and derangement in 1693.[54]

The end of the seventeenth century marked a transition from the esoteric alchemical definitions of "fermentation" to that of a mechanical process akin to the one described in *Meteorologica* 4. During the eighteenth century, Newton's theory of intercorpuscular attraction analogous to gravity gained considerable popularity.[55] An "example of the application of the theory to the problem of the nature of fermentation is provided by the writings of John Freind (1675–1728):

> The Fermentation we here undertake to Explain, is that Intestine Motion of Parts, which arises upon the Dissolution of Solids in *Liquors* or *Menstruums* . . . This Motion therefore may very well be accounted for, from an *Attractive Force*, which is so very extensive in *Natural Philosophy* . . . If this Motion increases to a very high degree, it raises an *Effervescency* and *Heat*, which is nothing else but a more rapid Motion of Parts . . . That this *Fermentation* is raised by *Elastick Particles*, is very probable, because all Bodies *ferment* more slowly, when debarred from the Air, which all allow is *Elastick*. So that to make *Ale ferment* well, we mix it with *Yeast*; a *Ferment* which abounds with *Air*.[56]

Before Freind, John Keill (1671–1721) was the first to apply Newton's ideas to chemical phenomena.

Stephen Hales (1677–1761), an important successor of Mayow in the study of "elastick airs," invented a "pneumatic trough" which enabled him to detect the release or uptake of gases from various materials when they were heated or underwent fermentation.[57] He considered that his experiments supported Newton's statement in Query 30 of the *Opticks* that "Dense bodies by Fermentation rarify into several sorts of Air, and this Air by Fermentation, and sometimes without it, returns into dense Bodies."[58]

---

[54] Golinski, J. (1988). "The secret life of an alchemist" in: *Let Newton Be!* (note 5), pp. 147–167.

[55] Spargo, P. E. and C. A. Pounds (1979). "Newton's 'derangement of the intellect.' New light on an old problem," *Notes and Records of the Royal Society of London* 34, pp. 11–32.

[56] Schofield, R. E. (1970). *Mechanism and Materialism. British Natural Philosophy in an Age of Reason.* Princeton University Press; Thackray, A. (1970). *Atoms and Powers: An Essay on Newtonian Matter-Theory and the Development of Chemistry.* Cambridge, Mass.: Harvard University Press.

[57] Freind, J. (1712). *Chymical Lectures*, pp. 70ff. London: Bawyer.

[58] Parascandola, J. and A. J. Ihde (1969). "History of the pneumatic trough," *Isis* 60, pp. 351–361.

Another Newtonian was Herman Boerhaave (1668–1738), the professor of medicine, botany, and chemistry at Leiden.[59] His lectures on chemistry attracted many students from abroad, and the authorized textbooks based on these lectures and demonstrations were outstanding. He devoted much attention to fermentation, and I quote several lengthy excerpts:

> I say then, that in every Fermentation, there is an intestine motion of the whole Mass, and all the parts, so long as this physical action continues; and I call it an intestine one, because it chiefly depends upon the internal principles of the vegetable Substances that are fermenting ... But I add further, that this intestine motion can be excited only in vegetable Substances ... I know very well, that some famous Authors make no scruple to assert the contrary; and therefore to distinguish here as nicely as possible, I define a true and perfect Fermentation by its proper effect, and that is, that always terminates in the production of either the Spirit or Acid ... Putrefaction is quite different from every Fermentation, for I cannot allow any thing to come under this name which don't either generate inflammable Spirits, or an Acid. For the same reason therefore all the various kinds of effervescences ... must be absolutely excluded likewise, though these properly come under the title of intestine Motions, and are often observed even in pure, vegetable Substances, as we see in very strong Vinegar, and fixed alkaline salt.[60]

Boerhaave listed several materials as ferments. Among them were:

> The Yeast, or fresh flowers of Malt Liquor, or Wine, which are thrown up to the top whilst they are in the action of Fermentation, for if his light, frothy Matter is mix'd with other fermentable Substances it wonderfully promotes their Fermentation, provided these Flowers are fresh, and not fallen ... The same Matter, afterwards grown heavier, and subsided to the bottom, if it is not too old ... The acid, mealy, fermented Dough or Leaven of the Bakers. For if fresh, sweet, wheaten Flower is kept in a dry place, and secured from Insects, it may be preserved for years without Corruption, but if this be kneaded with Water into a soft, stiff, sweet Dough, and this is lightly covered in a warm place, it begins the space of an Hour to grow lighter, puff up, and be full of Bladders, and lose its Smell, Taste, and Tenacity, and afterwards acquires both a sour Smell and Taste, which was then called *zyme, Fermentum*, a Ferment, and gave the first name to the whole

---

[59] Hales, S. (1727). *Vegetable Staticks*, p. 166. London: Innis and Woodward.

[60] Lindeboom, G. A. (1968). *Herman Boerhaave. The Man and his Work.* London: Methuen.

> Operation, for if this Leaven is mixed with fresh Dough not yet fermented, it will make it ferment much sooner, and more efficaciously than it would do otherwise. Hence then we see, that a Ferment may be soon prepared from a Body in which no Ferment actually existed before.[61]

In his personal chemical researches, Boerhaave studied the solubility of air in various liquids, confirmed Hales' findings on the release of "air" in fermentation, and offered a theory of an all-pervasive fire (*ignis pabulum*) which was assimilated into the Newtonian tradition.[62]

During the seventeenth century it became fashionable to use as models for acids particles with sharp spicules which fit into the pores of variously shaped earths. The transition is evident in a comparison of the two leading successive French textbooks of the time, those of Nicaise le Febure (also Le Fèvre, ca. 1610–1669) and Nicolas Lemery (1645–1715). Le Febure assumed the existence of a Paracelsian "universal spirit":

> And as this Spirit is universal, so can not be specificated but by means of particular Ferments, which do print in it the Character and Idea of mixt bodies, to be made such or such determinate substances, according to the diversity of Matrixes, which receive this spirit in themselves to make it a body.[63]

"Digestion" is defined as "one of he principal and most necessary Operations of Chymistry; because Mixts are made tractable by it, and capable to yield what we desire out of them."[64]

Lemery, whose book owes much to that of Le Febure, adopted the spicule-pore model, and defined "fermentation" as

> an effervescence caused by the spirits seeking to leave some body and encountering earthy and coarse particles which oppose their passage, cause swelling and rarefaction of the matter until they are detached. But in this detachment, the spirits divide, are subtilized and release the principles so that they transform the matter into a nature that it

[61] Boerhaave, H. (1735). *Elements of Chemistry* (translated by T. Dallowe).Vol. 2, p. 115. London: Pemberton et al.

[62] *Ibid.*, pp. 119–120.

[63] Metzger, H. (1930). *Newton, Stahl, Boerhaave et la Doctrine Chimique*. Paris: Alcan; Kerker, M. (1955). "Herman Boerhaave and the development of pneumatic chemistry," *Isis* 46, pp. 36–49; Heimann, P. M. (1973). "'Nature is a perpetual worker': Newton's aether and eighteenth-century natural philosophy," *Ambix* 20, pp. 1–25.

[64] Le Febure, N. (1670). *A Compleat Body of Chemistry*, p. 16. London: Pulleyn.

did not have previously. Although there is some difference between fermentation and effervescence, one does not scruple to take one for the other.[65]

Lemery, as a strict Cartesian, did not accept Van Helmont's finding of acid in the stomach: "There is no need to search for imaginary acids for digestion: the saliva which mixes with the food, to the extent one gives the first trituration with the teeth, is enough to excite the fermentation in the stomach."[66]

In 1690, Johann (Jean) Bernoulli (1663–1748), one of the greatest mathematicians of the eighteenth century, submitted a dissertation on effervescence and fermentation for the M.D. degree at the University of Basle. This delightful document has recently become available with an English translation.[67] Like Lemery, Bernoulli considered fermentation to be the same as effervescence, and his treatment dealt solely with the latter. He used as models a tetrahedron for what he called the "agent" and a stellate figure for the "patient", corresponding to an acid an alkali respectively. As he described it,

> . . . the bodies which seem to be fermented without addition of anything, such as must, barley macerated in water and others of the Same kind contain only particles which I called above the patient body. Therefore, to produce fermentation, particles of acid or agent body must arrive from the external or ambient air. Different experiments show that air is rich in acid particles. Therefore, when must or another fermentable substance is exposed to air, particles of acid introduce themselves little by little into the pores of must and combine with its alkaline particles which they disrupt in the way described above and enable the exit of the included air.[68]

A similar definition of fermentation was offered by George Wilson (1631–1711):

> *Fermentation*, is an Ebullition raised by Spirits that endeavour to separate themselves from the Body, but meeting with Earthy Parts that oppose their Passage, they swell, and rarify the liquor 'till they find

---

[65] *Ibid.*, p. 73.

[66] Lemery, N. (1701). *Cours de Chymie.* 9th ed., pp. 61–62. Paris: Delespine. See Bougard, M. (1999). *La Chimie de Nicolas Lemery*, pp. 209–213. Turnhout: Brepols.

[67] Bougard, M. (1999) (note 131), p. 282.

[68] Bernouilli, J. (1997). *Dissertations on the Mechanics of Effervescence and Fermentation and on the Mechanics of the Movement of Muscles.* Philadelphia: American Philosophical Society.

> their way out: In this Separation of Parts, the Spirits divide in such a manner, as to make the Matter of another Nature than it was before.[69]

To this must be added that in 1713, the physician Marcus Gerbezius (1658–1718) wrote that, in his view, alcoholic and acetic fermentation were chemical processes induced by certain miniscule particles, "volatile bodies," which escape from organic matter and are found suspended in the atmosphere.[70]

Another of the many contributors, during 1650–1700, of ideas about fermentation was Johann Joachim Becher (1635–1682).[71] A clever and energetic German, he rose to be a court physician in Mainz and Munich, and commercial advisor to Emperor Ferdinand III. Becher had a laboratory in Munich, and published several chemical works, the one known as *Physicae Subterraneae* (1667) the most famous one. He also gained a considerable reputation as a writer on economics.

Becher sought to clarify the use of the word *fermentatio* by distinguishing between three kinds of fermentation: that accompanied by the evolution of gas (*intumefactio*), the alcoholic fermentation of sweet liquors (*fermentatio proprie*), and the fermentation leading to the production of acid (*acetifactio*). He also believed that fermentation is akin to combustion, air is needed in the process, and that alcohol is not present in the original must of wine. Becher rejected the four elements of Aristotle and the three principles of Paracelsus, proposed a theory of three earths (vitreous, combustible, fluid) which correspond to the alchemical salt, sulfur, and mercury. He introduced the Greek word Φλόγιστόν (*phlogiston*) to denote the combustible earth sulfur. Becher also distinguished between simple bodies, compounds, and mixts. He suggested that because of the great complexity of animal matter, it is dissolved by putrefaction, while the lesser complexity of plant matter is dissolved by fermentation, and the simplest mineral form is dissolved by fusion (*liquefactio*).[72]

---

[69] *Ibid.*, p. 57.

[70] Wilson, G. (1700). *A Compleat Course of Chymistry*, p. 6. London: W. Turner.

[71] Grmek, M. (1972). "Gerbezius, Marcus". *Dictionary of Scientific Biography*, vol. 5, pp. 366–367. New York: Scribner's.

[72] Partington, J. R. (1961), pp. 637–652; Smith, P. H. (1994). *The Business of Alchemy. Science and Culture in the Holy Roman Empire*. Princeton University Press.

Georg Ernst Stahl (1660–1734), professor of medicine at Halle,[73] dismissed the Cartesian and Newtonian explanations of chemical phenomena and in his search for new "principles" he resembled the Paracelsians.[74] He took from Becher the idea of several types of fermentation, and in his *Zymotechnia Fundamentalis* defined the process as

> ... a colliding and rubbing motion, through an aqueous fluid, of very numerous molecules compounded (not very intimately or firmly) from Salt, Oil, and Earth. By the motion the bond of these principles is gradually loosened, and the principles are moved apart in the process, and attenuated by frequent rubbing.[75]

Stahl thought that a putrefying substance can transfer its internal motion to aquiescent substance if it is disposed to such an inner movement. Like Lemery, Stahl rejected Van Helmont's claim to have found a gastric ferment. Stahl also attempted to clarify the concepts of elements, compounds and aggregates:[76]

> All natural Bodies are either simple or compounded; the simple do not consist of physical parts; but the compounded do. The simple are Principles, or the first material causes of Mixts; and the compounded, according to the difference of their mixture, are either mix'd, compound, or aggregate; mix'd if composed merely of principles; compound, if form'd of Mixts in any determinable single thing; and aggregate, when several such things form any other entire parcel of matter, whatever it be.[77]

Stahl's phlogiston, which he invoked for fermentation as well as combustion, was accepted by several of the most productive chemists of the century, including Andreas Marggraf, Joseph Black, and Carl

---

[73] Partington, J. R. (1961), pp. 653–686; King, L. S. (1975). *Dictionary of Scientific Biography* 12, pp. 599–606. New York: Scribner's.; Berger, J. (2000), "Atomismus und 'vernunftige chemische Erfahrung': Grundlage der chemischen Materietheorie Georg Ernst Stahls," *Nova Acta Leopoldina* 30, pp. 125–143; Ströker, E. (2000). "Georg Ernst Stahls Beitrag zur Grundlegung der chemischen Wissenschaft," *ibid.*, pp. 145–160.

[74] King, L. S. (1964). "Stahl and Hohann: A study in eighteenth-century animism," *Journal of the History of Medicine* 19, pp. 118–130.

[75] Quoted from Chang, K. (2002). "Fermentation, phlogiston, and matter theory: Chemistry and natural philosophy in Georg Ernst Stahl's *Zymotechnia Fundamentalis*," *Early Science and Medicine* 7, pp. 31–64 (38).

[76] Oldroyd, D. (1973). "An examination of G. E. Stahl's *Philosophical Principles of Universal Chemistry*," *Ambix* 20, pp. 36–52. See Metzger, H. (1930). *Newton, Stahl, Boerhaave et la Doctrine Chimique.* Paris: Alcan.

[77] Quoted from Oldroyd, D. (1973), p. 43.

Scheele.[78] Stahl also achieved some fame for his vitalist view that the human body would undergo putrefaction if it were not organized and protected by the soul (*anima*). This doctrine, expounded in his book *Theoria Medica Vera* (1708), was criticized by Gottfried Leibniz (1646–1716):

> The distinguished author rightly says . . . that chemistry seems still more distant from the aim of the physician than anatomy. Yet I should wish that not even it be too far removed. For although different acids, bases and oils have very different effects still they have much in common, the observation of which paves the way to more pertinent matters. Changes in animals are certainly very different from changes in plants, and there is perhaps nothing in our body that corresponds in the strict sense to fermentation, through which plants are fitted to produce alcohol and finally acid, yet in animals there is a certain proper chemistry, so to speak, and changes that take place in the humors of animals belong no less to chemistry than those occurring in the fluids of plants.[79]

Until the beginning of its demise during the 1780s, the phlogiston theory was widely accepted (often in revised form) by European chemists. In France, the noted lecturer Guillaume François Rouelle (1703–1770) and his pupil Pierre Jacques Macquer (1718–1784) were the leading supporters of the theory.[80] The successive editions of Macquer's textbook defined fermentation as

> an intestine motion, which, arising spontaneously among the insensible parts of a body, produces a new disposition and a different combination of those parts. To cause a fermentation in a mixt body, it is necessary, first, that there be in the composition of that mixt a certain proportion of watery, saline, oily, and earthy parts . . . Secondly, it is requisite that the body to be fermented be placed in a certain degree of temperate heat. Lastly, the concurrence of the air is also necessary to fermentation.[81]

---

[78] Partington, J. R. and D. McKie (1937–1939). "Historical Studies on the phlogiston theory," *Ambix* 2, pp. 361–404; 3, pp. 1–53, 337–371; 4, pp. 113–149.

[79] Quoted from Rather, L. J. and J. B. Frerichs (1968). "The Leibniz-Stahl controversy—I. Leibniz' opening objections to the *Theoria medica vera*," *Clio Medica* 3, pp. 21–40 (32). See Peters, H. (1916). "Leibniz als Chemiker," *Archiv für die Geschichte der Naturwissenschaften und der Technik* 7, pp. 87–106, 220–234, 275–287.

[80] Rappaport, R. (1961). "G. F. Rouelle: An eighteenth-century chemist and teacher," *Chymia* 6, pp. 68–101; (1962). "Rouelle and Stahl—The phlogistic revolution in France," *Chymia* 7, pp. 73–102; Fichman, M. (1971). "French Stahlism and chemical studies of air," *Ambix* 18, pp. 94–122.

[81] Macquer, P. J. (1777). *Elements of the Theory and Practice of Chemistry*. 5th ed., pp. 83–101. Edinburgh: Donaldson and Elliot. See Coleby, L. J. M. (1938). *The chemical studies of P. J. Macquer*. London: Allen and Unwin.

Like Macquer, Jacques François Demachy (1728–1803), defined fermentation as an intestine movement, but disagreed about the participation of air in the process. He believed that the "pellicule which forms on the surface of fermenting liquids is able to penetrate the thinner portions, and in absorbing their motion becomes able to determine and accelerate the fermentative motion... This material is called the yeast or ferment."[82]

In 1754, the problem of the nature of vinous fermentation, and the accompanying effervescence assumed a different aspect. In that year, Joseph Black (1728–1799) presented at the University of Edinburgh his Latin M.D. dissertation, and in the following year published a revised English translation which is counted among the classics of experimental chemistry.[83]

Black showed that quicklime, which is very caustic, absorbs an "air" to form a mildly alkaline substance, and when chalk is roasted, this "air" is released and gives a precipitate with limewater. He named this component of common air "fixed air."[84] Fermentation is not mentioned in the 1755 paper, but in his lectures to students he is reported to have said that

> in 1757 he had found that fixed air is the chief part of the elastic matter which is formed in the vinous fermentation. Van Helmont had indeed said this, and it was to this that he gave the name *gas sylvestre* ... I convinced myself of the fact by going to a brew-house with two phials, one filled with distilled water, and the other with lime-water. I emptied the first into a vat wort fermenting briskly, holding the mouth of the phial close to the surface of the wort. I then poured some of the lime-water into it, shut it with my finger, and shook it. The lime-water became turbid immediately.[85]

Joseph Priestley (1733–1804) made a similar visit to a brewery in about 1772.

---

[82] Demachy, J. F. (1766). *Instituts de Chymie*, vol. 1, pp. 264–273. Paris: Lottin.

[83] Black, J. (1944). *Experiments upon Magnesia Alba, Quicklime, and some other Alcaline Substances*. Edinburgh: Oliver and Boyd (Alembic Club Reprint No. 1).

[84] Guerlac, H. (1957). "Joseph Black and Fixed Air. A bicentenary retrospective, with some new or little known material," *Isis* 48, pp. 124–151, 433–456. See Donovan, A. L. (1975). *Philosophical Chemistry in the Scottish Enlightenment*. Edinburgh University Press; Breathnach, C. S. (2000). "Joseph Black (1728–1799): an early adept in quantification and interpretation," *Journal of Scientific Biography* 8,149–155.

[85] Black, J. (1803). *Lectures on the Elements of Chemistry*, J. Robison (ed.), p. 88. Edinburgh: Longman & Rees London and Creech Edinburgh.

The properties of fixed air and its formation during vinous fermentation were described in 1766 by Henry Cavendish (1731–1810).[86] He defined fixed air as "that species of factitious air, which is produced from alkaline substances, by solution in acids, or by calcination and showed that the air produced in the fermentation of sugar or apple juice had the same density and solubility in water as that produced by the action of acids on marble. By determining the loss of weight of the marble, he obtained a value of 40.7 per cent of fixed air (the correct value is 44 per cent). In like manner, he found a value of 57 per cent for the liberated fixed air upon the fermentation of brown sugar (too high, possibly due to evaporation of some alcohol). Aqueous solutions of fixed air were acidic, and it was proposed that it be designated the "universal acid;"[87] it was renamed "acide carbonique" by Lavoisier.

In 1752, before the publication of Black's famous paper, the army physician John Pringle (1707–1782), in writing about his work on fermentation and putrefaction, wrote about the word "ferment":

> It were to be wished, to avoid ambiguity, that we had two different words to denote the exciting cause of these intestine motions: but this is the less to be expected, on account of the disposition of all putrid animal substances to promote both animal putrefaction and a vinous fermentation in vegetables, as will appear by the sequel of these experiments.[88]

After the discovery of fixed air, the physician David Macbride (1726–1778) continued Pringle's work, and with an apparatus devised by Black showed that an alimentary mixture of meat, bread and water emits fixed air, which appears to inhibit the putrefaction (as judged by the "sweet" odor). According to Macbride, "Now since it appears, that these mixtures ferment so very quickly, even when unassisted by heat, how can there be any doubt that they must run

---

[86] Berry, A. J. (1960), *Henry Cavendish.* London: Hutchinson; McCormmach, R. (1961). "Henry Cavendish: A study of rational empiricism in eighteenth-century natural philosophy," *Isis* 60, pp. 293–306.

[87] Cavendish, H. (1921). *The Scientific Papers*, E. Thorpe (ed.). Vol. 2, pp. 96–101. Cambridge University Press; Le Grand, H. E. (1973). "A note on fixed air: the universal acid," *Ambix* 20, pp. 88–94.

[88] Quoted from Scott, E. L. (1970). "The 'Macbridean doctrine' of air: An eighteenth-century explanation of some biochemical processes, including photosynthesis," *Ambix* 17, pp. 43–57.

through the same process when they are received into the warm stomach, and are put in motion by the fermentative power of the saliva?[89]

Macbride's statement echoes the view of Jean Astruc (1684–1766) who argued against the role of trituration in digestion but also rejected Van Helmont's claim for a gastric ferment, with the saliva, bile, and pancreatic juice as the sole agents in the digestion of nutrients.[90] During the 1750s, René Antoine Ferchaut de Reaumur (1683–1757) passed metal tubes containing meat into the stomachs of birds, and showed that the meat was dissolved. This approach was extended by Lazzaro Spallanzani (1729–1799), thus establishing the existence of a gastric ferment.[91]

The period covered in this chapter witnessed the replacement of the mystical Neoplatonic view of fermentation evident in the writings of Paracelsus and Van Helmont by a Newtonian mechanism in corpuscular attraction and repulsion. There was also the emergence of a distinctive chemical philosophy envisioned by Van Helmont, who performed quantitative chemical experiments, but who also wrote about ferments which made metals from water. His volatile "wild spirit" produced from burning charcoal anticipated the discovery by Joseph Black (1728–1799) of "fixed air" produced during combustion or fermentation. As will be seen in the next chapter this discovery spurred the search for other volatile components of common air, long considered to be a homogeneous substance. Van Helmont's had a considerable influence on the chemical thought of Robert Boyle and of several English physicians, notably Edward Jorden and Thomas Willis. To these men may be added John Mayow, who interpreted his experimental results on respiration and combustion in terms of the presence of a "nitro-aerial spirit."

The cause of a distinctive chemical philosophy was also markedly promoted by Georg Ernst Stahl (1660–1734) who regarded chemistry as an independent discipline, with its own methods and con-

---

[89] Macbride, D. (1764). *Experimental Essays on the Following Subjects: I. On the Fermentation of Alimentary Mixtures*, p. 16. London: A. Miller.

[90] Astruc, J. (1711). *Mémoire sur la cause de la Digestion des Alimens*. Montpellier: Honoré Pech.

[91] Reaumur, R. A. F. de (1761). *Sur la Digestion des Oiseaux. Second memoire*. Amsterdam: Schreuder et Mortier; Spallanzani, L. (1789). *Dissertations relative to the Natural History of Animals and Vegetables*. 2 vols. London: J. Murray. See Friedman, H. C. (ed.) (1981). *Enzymes*, pp. 24–71. Stroudsburg, Pennsylvania: Hutchinson Ross.

cepts. He adopted Becher's idea of Phlogiston, which was widely accepted by leading chemists during most of the eighteenth century, but rejected by Lavoisier, who replaced it with the short-lived concept of caloric.

Around 1700, the effervescence was considered by some adherents of the mechanical philosophy to be the most significant characteristic of fermentation. John Mayow and Johann Bernoulli are notable examples. They offered mechanistic theories of the kind described above for Bernoulli, a member of a noted family of mathematicians. After his 1690 dissertations *On the Mechanics of Effervescence and Fermentation* and *On the Mechanics of the Movement of Muscles*, Johann Bernoulli became professor of mathematics in Groningen and in 1705 he succeeded his deceased brother Jacob as professor of mathematics in Basel.

CHAPTER THREE

LAVOISIER TO FISCHER

The first recorded evidence of Antoine Lavoisier's (1743–1794) interest in the problem of fermentation was the following notation in his laboratory notebook on February 20, 1773:

> Before commencing the long series of experiments that I propose to make on the elastic fluid which is released from bodies, whether in fermentation, or distillation, or finally by every type of combination, as well as [on] the air absorbed in the combustion of a great number of substances, I believe that I ought to put some reflections here in writing, in order to shape for myself the plan which I must follow.[1]

In 1774, Lavoisier published his *Opuscules Physiques et Chimiques.* Nearly a half of the book was devoted to summaries of the studies on "fluides elastiques" by investigators from Hales to Priestley, and the rest reported on Lavoisier's own work, including the repetition of experiments of others.[2] Much of his work during 1773 dealt with the question of whether fixed air is the "air" absorbed during the calcination of sulfur or phosphorus, and of metals such as lead or mercury. At one stage, he thought that the "acidum pingue" described by Johann Friedrich Meyer (1705–1765), who opposed Black's theory of causticity, might be the elastic air he was hoping to find, and he also examined the "nitrous air" of Joseph Priestley (1733–1804) as another possibility. In November 1772, Lavoisier had deposited with the secretary of the *Académie des Sciences* a sealed note in which he presented his theory that an elastic air is taken up in the calcination of metals and other substances. The note was opened in May 1773, and ended with the statement that "this discovery seeming to me one of the most interesting made since the time of Stahl, I thought it my

---

[1] Quoted from Holmes, F. L. (1985). *Lavoisier and the Chemistry of Life*, p. 7. Madison, Wis.: University of Wisconsin Press.

[2] Lavoisier, A. (1774). *Opuscules Physiques et Chimiques*. Paris: Deterville. Annotated English translation (1776) by Thomas Henry: *Essays Physical and Chemical.* London: Joseph Johnson.

duty to assure myself of priority by depositing this note.[3] At the end of 1773, however, the status of Lavoisier's elastic air in relation to Black's fixed air or Meyer's acidum pingue was unclear. According to Frederic L. Holmes:

> By January 1774, Lavoisier had learned some hard lessons. The beautiful rigorous style of demonstration that he admired in geometry would not work in chemistry. The many setbacks he had encountered as he tried to gather evidence to support a brave new theory had taught him that he, too, must follow "another route."[4]

In the preface to the *Opuscules*, Lavoisier stated that "I have also deferred the publication of my experiments on fermentation in general, and on the acid fermentation in particular."[5] His early interest is indicated in an unpublished manuscript of 1773, in relation to the fixation of air in fermentation:

> This absorption of surplus air is the same in the formation of all the acids. In the fermentation of beer wort . . . it is observed that a very great abundance of air is released as soon as the spiritous fermentation begins. But when in the progress of fermentation the liquor begins to turn to acid, soon all the air that was released is re-absorbed to enter into the composition of the acid. I have observed this phenomenon of absorption of air in every souring liquor.
>
> M. Abbé Rozier in his treatise on wine was the first to be struck by this phenomenon . . . It is easy to sense that these experiments must inevitably lead to a completely new theory of fermentation.[6]

He did not attempt a quantitative accounting of the chemical process in vinous fermentation until 1786–1789,[7] after he had gained great renown for his chemical contributions. Apart from his role in the rediscovery of what came to be called "oxygen," Lavoisier's achievements included his studies on combustion and respiration, the composition of nitric acid, the nature of heat, the synthesis of water, oxygen as the acidifying principle, saltpeter, the replacement of the

[3] Quoted from Kohler, R. E., Jr. (1972). "The origin of Lavoisier's first experiments on combustion," *Isis* 63, pp. 349–355 (351).

[4] Partington, J. R. (1962). *A History of Chemistry*, vol. 3, p. 385. London: Macmillan.

[5] Holmes, F. L. (1998). *Antoine Lavoisier—The Next Crucial Year*, p. 139. Princeton University Press.

[6] Lavoisier, A. (1776), p. xx.

[7] Daumas, M. (1955). *Lavoisier Théoricien et Expérimenteur*, pp. 59–63. Paris: Presses Universitaires de France.

weightless phlogiston in charcoal by the weightless caloric in oxygen gas, and his participation in the reform of the chemical nomenclature.

In his famous *Traité Élémentaire de Chimie*, Lavoisier devotes a chapter to one of his experiments, in which cane sugar was converted to alcohol and carbonic acid gas in the presence of brewer's yeast:

> This process (*opération*) is one of the most striking and extraordinary of all those presented to us by chemistry, and we have to examine whence comes the carbonic acid gas which is released, and how a sweet body, a vegetable oxide, can transform itself into two so different substances, one of which is combustible, and the other eminently incombustible. One sees that to solve these two questions, it is first necessary to know well the analysis and nature of the fermentable body, and the products of the fermentation; since nothing creates itself, neither in artificial operations nor those of nature, and one can take it for granted (*poser*) that in all operations, there is same quantity of matter before and after the operation; that the quantity and quality of the principles is the same, and that there are only changes, modifications.
>
> It is on this principle that the whole art of making experiments in chemistry is founded. One must suppose in every case a true equality or equation between the principles of the bodies one examines, and those which obtains through the analysis. Thus, since the grape must yield carbonic acid gas & alkohol (sic), I can say that *grape must* = *carbonic acid* + *alkohol*. It follows that one can arrive in two ways at a clarification of what happens in vinous fermentation; first, by determining carefully the nature and principles of the fermentable body; second, by observing carefully the products which result from the fermentation, & it is evident that the knowledge one can acquire about the one will lead to certain conclusions about the other.[8]

In considering the changes in elementary composition in the conversion of sugar to alcohol and carbonic acid, Lavoisier first thought that "the *matière charbonneuse* contained in sugar decomposes water, forming fixed air with the oxygen principle and releasing the inflammable air to form, somehow, the spirit of wine."[9] In the *Traité*, however, he concluded that

> The effect of the vinous fermentation is thus reduced to separating the sugar, which is an oxide into two portions; one part is oxygenated at the expense of the other, so as to form carbonic acid; to deoxygenat-

[8] Lavoisier, A. (1789). *Traité Élémentaire de Chimie*, pp. 140–141. Paris: Cuchet. See Siegfried, R. (1989). "Lavoisier and the conservation principle," *Bulletin for the History of Chemistry* 5, pp. 18–24.

[9] Holmes, F. L. (1985), p. 292.

> ing the other part in the favor of the first, to form a combustible substance which is *alkool* (sic); therefore, if it were possible to reunite these two substances, *alkool* and carbonic acid, one would reform sugar . . . I had formally advanced in my first papers on the formation of water, that this substance regarded as an element, is decomposed in numerous chemical operations, notably in vinous fermentation; I then supposed that water was present in sugar, while I am now persuaded that it only contains the materials for its formation. One can imagine how much it cost me to abandon my first ideas; it is only after several years of reflection, and after a long series of experiments and observations on vegetables that I am persuaded by it.[10]

Lavoisier considered the carbon, oxygen, and hydrogen in sugar to be combined in such a way that a slight force is sufficient to disturb the equilibrium in their connection.

For some historians of science, Lavoisier's affirmation of the law of the conservation of weight and his use of balance sheets were a more important theoretical feature of the fermentation experiment than his ideas about the nature of the chemical process.[11] Apart from the conservation of matter, a more specific assumption was that there were no products other than alcohol, carbonic acid, and acetic acid. A remarkable feature of the balance sheet is the faithful reproduction of the total weight (400 lbs. water + 100 lbs. sugar + 10 lbs. yeast = 510 lbs.) in tables for the analytical data for the hydrogen, oxygen, and carbon content of the initial components of the fermentation mixture and of the final products (including acetic acid) and unfermented sugar. Lavoisier introduced a combustion method for the determination of the elementary composition of sugar, which he considered to be composed of carbon and water. He reported (by weight) 64% oxygen, 28% carbon, and 8% hydrogen. In 1811, however, Jacques Louis Gay-Lussac (1778–1850) and Louis Jacques Thenard (1777–1857) developed an improved method of combustion analysis and found for sucrose 50.63% oxygen, 42.47% carbon, and 4.9% hydrogen. As Thenard described it:

---

[10] Lavoisier, A. (1789), pp. 150–151.

[11] Freund, I. (1904). *The Study of Chemical Composition*, pp. 58–63. Cambridge University Press; Siegfried, R. (1989). "Lavoisier and the conservation of weight," *Bulletin for the History of Chemistry* 5, pp. 18–24; Holmes, F. L. (1994). "Lavoisier—The conservation of matter," *Chemical & Engineering News* September 12, pp. 38–45.

> Lavoisier endeavoured to accomplish the analysis of organized bodies by burning them in oxygen gas, and MM. Gay-Lussac and Thenard have lately proposed a different method of examining vegetable and animal substances, which consists in changing them into water, carbonic acid and azote by combustion by means of potassium chlorate.[12]

In 1815, Gay-Lussac reported that he had checked and corrected the figures given by Lavoisier, and that, from 100 parts of sugar 51.34 had been converted into alcohol and 48.68 into carbonic acid; this was interpreted to indicate that sugar had been converted into equal parts of the two products.[13] Gay-Lussac's equation for alcoholic fermentation is usually written as

$$C_6H_{12}O_6 = 2\ CO_2 + 2\ CH_3CH_2OH$$

In recognizing the originality of Lavoisier's analytical method, and the stimulus it provided for its improvement, one must agree that all the quantitative data for the fermentation experiment are highly suspect, and "must be regarded as one of those remarkable instances in which the genius of the investigator triumphs over experimental deficiencies, for the analytical numbers employed contained grave errors, and it was only by a fortunate compensation of these that a result so near the truth was attained."[14] In Holmes's judgment

> It is clear that he sought reliability, but not great precision. He routinely estimated the magnitude of errors due to small losses he could not measure. He aimed for a complete balance of all the materials before and after an operation, but did not expect to arrive at measured quantities exactly. If they came close enough to support his interpretation of the operation he was studying, that was good enough for him.[15]

Lavoisier was a victim of the Reign of Terror, because of his membership in the *Ferme*, a private financial consortium charged by the royal government to handle leases and to collect taxes. It is certain that if he had been spared, he would have continued his fermenta-

---

[12] Thenard, J. L. (1819). *An Essay on Chemical Analysis* (translated by J. G. Children), p. 350. London: W. Phillips.

[13] Gay-Lussac, J. L. (1815). "Lettre à M. Clément, sur l'analyse de l'alcohol et de l'éther sulfurique, et sur les produits de la fermentation," *Annales de Chimie* 95, pp. 311–318.

[14] Harden, A. (1923). *Alcoholic Fermentation.* 3rd ed., pp. 2–3; Partington, J. R. (1962), p. 480.

[15] Holmes, F. L. (1998) (note 161), p. 89.

tion experiments with new, but unused, apparatus he had acquired for his laboratory.[16]

One of Lavoisier's partners in the reform of the chemical nomenclature, Louis Bernard Guyton de Morveau (1737–1816), defined fermentation as "spontaneous intestine motion which destroys the organization of bodies, separates their principles, and arranges them in new combinations, from which there results another compound with totally different properties."[17]

The new anti-phlogistic chemistry found expression in the voluminous writings of Antoine François de Fourcroy (1755–1809). In his *Philosophie Chimique*, the chapter on fermentation began as follows:

> After plants and animals are deprived of life, or when their products are removed from the individuals of which they formed a part, there ensues in them motions which destroy their tissue and alter their composition. These motions constitute the various kinds of fermentation. In exciting them nature's aim is manifestly to convert the compounds made by vegetation and animalization into simpler substances, and to make them enter into new combinations of different kinds.[18]

According to Fourcroy, the cause of vinous fermentation "appears to be due to the decomposition of sugar, a large part of its oxygen taken up by carbon, burning it and converting it into carbonic acid. At the same time the hydrogen, remaining in the decarbonized sugar and combining with it, gives rise to alcohol."[19]

Fourcroy also called attention to the 1787 paper of Adamo Fabbroni (1752–1822) in which the latter claimed that "fermentation is only a decomposition of one substance by another, like that of a carbonate by an acid, or of sugar by nitric acid . . . The material which decomposes sugar in vinous effervescence is the vegeto-animal substance."[20] This substance had been discovered in 1728 by Jacopo Bartolomeo Beccari (1682–1766) upon washing out the starch from

---

[16] Daumas, M. (1950). "Les appareils d'experimentation de Lavoisier," *Chymia* 3, pp. 45–62 (61).

[17] Guyton-Morveau, L. B. (1778). *Élémens de Chymie Théorique et Pratique*. Vol. 3, pp. 265–266. Dijon: Frantin.

[18] Fourcroy, A. F. (1806). *Philosophie Chimique*. 3rd ed., p. 357. Paris: Levrault, Schoell. See Smeaton, W. A. *Fourcroy Chemist and Revolutionary*. Cambridge: Heffer.

[19] *Ibid.*, p. 359.

[20] Fourcroy, A. F. (1799). "D'un mémoire du cit. Fabroni, sur les fermentations etc." *Annales de Chimie* 31, pp. 299–327 (301–302).

wheat flour with water, but not reported until twenty years later.[21] He named it "gluten vegetabile." Fabbroni's claim was taken up by Thenard, who considered the real ferment to be a nitrogenous material present in beer yeast and resembling coagulated albumin. Thenard also concluded that

> ... the ferment removes oxygen from the sugar, not only by means of a part of its carbon but also by means of part of its hydrogen, For the quantity of carbon given up by the ferment is too little to be the only gem of fermentation, nitrogen disappears and enters perhaps into the composition of the alcohol; the other principles of the ferment form acetic acid and a particular white material which precipitates.[22]

In 1810 there appeared a book by Nicholas Appert (1750–1851), a French manufacturer of confectionary, distilled spirits, and food products, in which he described methods for preserving foods by putting them into tightly sealed vessels that were then heated in boiling water. His success marked the beginning of the canning industry.[23] Gay-Lussac examined Appert's results and found that on opening the sealed containers

> These substances on contact with air promptly acquire the tendency for putrefaction or fermentation, but when they are submitted to the temperature of boiling water in well-closed vessels, the absorbed oxygen produces a new combination which is no longer able to excite putrefaction or fermentation, or which becomes coagulated by the heat in the same manner as albumin.[24]

According to Jean Antoine Chaptal (1756–1832), who became Bonaparte's minister of the interior (1800–1804),

> The necessary conditions for fermentation are, 1) contact of pure air, 2) a certain degree of heat, 3) a more or less considerable quantity of fermenting liquid ... The phenomena that essentially accompany fer-

[21] Bailey, C. H. (1941). "A translation of Beccari's lecture 'Concerning Grain' (1728)," *Cereal Chemistry* 18, pp. 555–561; Breach, E. F. (1961). "Beccari of Bologna, the discoverer of vegetable protein," *Journal of the History of Medicine* 16, pp. 354–373.

[22] Thenard, L. J. (1803). "Mémoire sur la fermentation vineuse," *Annales de Chimie* 46, pp. 294–320 (318–319).

[23] Bitting, A. W. (1937). *Appertizing or the Art of Canning: Its History and Development.* San Francisco: The Trade Pressroom.

[24] Gay-Lussac, J. L. (1810). "Mémoire sur la fermentation," *Annales de Chimie* 76, pp. 243–259 (255). See Crosland, M. (1978). *Gay-Lussac: Scientist and Bourgeois.* Cambridge University Press.

mentation are 1) the production of heat, 2) the absorption of oxygen gas. One can facilitate fermentation 1) by increasing the volume of the fermenting mass; 2) by using an appropriate leaven . . . One can distinguish two kinds of leaven: 1) eminently putrescible bodies whose addition hastens the fermentation; 2) those already containing oxygen, and which consequently furnish a greater amount of this principle to the fermentation.[25]

Chaptal also wrote: "If there is insufficient sweet substance, one may add sugar."[26] When he offered this advice to French wine makers with the prestige of his ministerial office, it became known as "chaptalization."[27]

In 1800, the Institut de France offered a one-kilogram gold medal for the best answer to the question: "What are the characteristics by which animal and vegetable substances which act as ferments can be distinguished from those which they are capable of fermenting?" The prize was offered again in 1804, but was never awarded.

In the German states, Johann Friedrich Westrumb (1732–1816), an apothecary in Hameln, called fermentation

> . . . a great and general operation of nature. Dissolution and decomposition of bodies into their near and distant constituents. Are the things that we obtain by fermentation really in the bodies from which we obtain them? Certainly . . . What is wine other than the previous grape juice, only now through the fermentation dissolved, refined, ennobled, if we wish to call it so, actually however separated into the distant constituents of the juice. Was not the constituent, though not the form, still like the parts? This is indeed the spirit, the noblest part of the wine, a sweet substance composed from tartaric acid, an oil refined in the highest degree (a plant combustible), and water.[28]

Westrumb also criticized the theory of another German apothecary Johann Christian Wiegleb (1732–1800), who supposed that alcohol and vinegar are present as such in the fermentable substance and are separated by fermentation.[29]

---

[25] Chaptal, J. A. (1796). *Élémens de Chymie.* 3rd ed. Vol. 3, pp. 275–277. Paris: Deterville.

[26] *Ibid.*, p. 280.

[27] Paul, H. W. (1996). *Science, Vine, and Wine in Modern France*, pp. 123–130. Cambridge University Press.

[28] Westrumb, J. F. (1788). *Kleine physikalisch-chemische Abhandlungen*, vol. 2, p. 273. Leipzig: Johann Gottfried Muller.

[29] Partington, J. R. (1962), p. 569.

Further evidence of the interest in the problem of fermentation was provided by the sizable number of M.D. dissertations at the University of Edinburgh during 1773–1790 which were entitled *De Fermentatione.*[30]

In 1813, little notice was taken of the pharmacist Charles Bernard Astier (1771–1837) who claimed that "the air is the vehicle of every kind of germs" and is the source of the ferment which ". . . lives and nourishes itself at the expense of the sugar, whereby there results a disruption of the equilibrium among the elementary units of the sugar."[31] Nor was much account given to the opinion of Christian Polykarp Friedrich Erxleben (1765–1831), a Bohemian pharmacist and brewer, who stated in 1818 that

> As a rule one cannot state in advance with certainty the result of any known chemical *operation*, but here there is an exception. Because the fermentation, although until now always considered as such, appears in no way a mere chemical *operation* but much rather in part a process by which plants grow, and must be considered as the link in the great chain in nature which brings about union of the activities we call chemical processes with those of plantlike growth.[32]

The demonstration that the agents of fermentation were living organisms came in 1837, when three investigators (Caignard de la Tour, Schwann, Kützing) independently, and almost simultaneously, reported their microscopic observations and experimental results. As so often in the history of a scientific problem, this instance of multiple discovery was an outgrowth of improvement in instrumentation, in this case the invention of the achromatic compound microscope. The beginnings of microscopy in the seventeenth century led to the instrument used by Antony van Leewenhoek (1632–1723), who saw spermatozoa, red blood corpuscles, and many kinds of protozoa and bacteria (which he called "little beasts"), as well as globules of yeast.

---

[30] Kendall, J. (1952). "The first chemical society, the first chemical journal, and the Chemical Revolution," *Proceedings of the Royal Society of Edinburgh* A63, pp. 346–358, 385–400; Perrin, C. E. (1982). "A reluctant catalyst: Joseph Black and the Edinburgh reception of Lavoisier's chemistry," *Ambix* 29, pp. 141–176 (172).

[31] Astier, C. B. (1813). "Expériences faites sur le sirop et le sucre de raisin," *Annales de Chimie* 87, pp. 271–285 (274).

[32] Erxleben, P. C. F. (1818). *Ueber Guete und Staerke des Bieres, etc.*, p. 69. Prague: Haase. Quoted from Teich, M. (1992). *A Documentary History of Biochemistry 1770–1940*, p. 16. Rutherford: Fairleigh Dickinson University Press.

Leewenhoek's microscope permitted magnifications of several hundred diameters, and was not matched with compound microscopes until after 1800. Before then, the limits of optical resolution were often exceeded by the imagination of the observers, and during the eighteenth century the instrument fell into disrepute.

Charles Caignard de la Tour (1779–1859) was a professor at the military school in Paris and a noted inventor. He presented his results initially in 1835–1836, then on 12 June 1837 before the *Académie des Sciences*, and published in 1838. His principal results were:

> 1. That the yeast of beer (this ferment of which one makes so much use and which reason was suitable for examination in a particular manner) is a mass of little globular bodies able to reproduce themselves, consequently organized, and not simply organic or chemical, as one supposed. 2. That these bodies appear to belong to the vegetable kingdom and to regenerate themselves in two different ways. 3. That they seem to act on a solution of sugar only as long as they are living. From which one can conclude that it is very probably by some effect of their vegetable nature that they disengage carbonic acid from this solution and convert it into a spirituous liquor.[33]

Theodor Schwann (1810–1882) was associated with the leading physiologist Johannes Müller (1801–1858) in Berlin when he published his report on alcoholic fermentation in 1837. In a remarkable surge of scientific productivity, in the previous year Schwann had written a lengthy paper on pepsin (to be discussed shortly), and in 1839 he presented the outlines of the cell theory of living organisms. In his microscopic observation of yeast, he saw the same budding globules as did Caignard de la Tour, and termed them *Zuckerpilz* ("sugar-fungus", later named *Saccharomyces*). He concluded that

> Vinous fermentation must be regarded as the decomposition effected by the sugar fungus, which extracts from the sugar and a nitrogenous substance the materials needed for its own nutrition and growth, and whereby such elements of these substances bodies (probably among others) as are not taken up by the fungus combine preferentially to form alcohol.[34]

---

[33] Caignard de la Tour, C. (1838). "Mémoire sur la fermentation vineuse," *Annales de Chimie* 68, pp. 206–222 (221).

[34] Schwann, T. (1837). "Vorläufige Mitteilung, betreffend Versuche über die Weingährung und Fäulnis," *Annalen der Physik* 41, pp. 184–193 (192).

Two years later, in the famous book on his cell theory, Schwann wrote:

> That this fungus is the cause of the fermentation follows, in the first place, from its constant occurrence in fermentation, secondly because the fermentation ceases under all conditions which visibly kill the fungus, namely boiling, treatment with potassium arsenite etc., thirdly because the exciting principle in the fermentation process must be a material that is evoked and increased by the process itself, a phenomenon that applies only to living organisms.[35]

Friedrich Traugott Kützing (1807–1893), first a pharmacist and after 1838 a science teacher at the Hochschule in Nordhausen, reached similar conclusions about brewer's yeast, and knew of the work of Caignard de la Tour and Schwann when he wrote: "Now that the three of us have made the same observations in regard to the truly organic nature of yeast, I am all the more happy that my findings were confirmed by other scientists. I therefore gladly renounce a claim to priority, since it does not matter for science who made the discovery first."[36]

The evidence presented in support of the organismic theory of alcoholic fermentation was confirmed by several investigators, notably Pierre Jean François Turpin (1775–1840), Théodore Auguste Quevenne (1805–1855), and Eilhard Mitscherlich (1794–1863).[37] Moreover, the theory was adopted by Karl Josef Balling (1805–1868) and Friedrich Wilhelm Lüdersdorff (1801–1886), two leading specialists in brewing technology. The challenge to the chemists received a quick response in an anonymous article in Liebig's *Annalen der Chemie* for 1838, in which yeast was elaborately described as a tiny animal shaped like a distilling flask; under the microscope this organism could be seen to swallow sugar, digest it in its stomach, and excrete alcohol through its digestive tract and carbonic acid through its urinary tract.[38]

---

[35] Schwann, T. (1839). *Mikroskopische Untersuchungen*, p. 235. Berlin: Sander.

[36] Kützing, F. G. (1837). "Mikroskopische Untersuchungen über die Hefe etc.," *Journal für prakische Chemie* 11, pp. 385–409 (386).

[37] Turpin, P. J. F. (1838). "Mémoire sur la cause et les effets de la fermentation alcoolique et aceteuse," *Comptes Rendus* 7, pp. 369–402; Quevenne, T. A. (1938). "Étude microscopique et chimique du Ferment, suivie d'expériences sur la fermentation alcoolique," *Journal de Pharmacie* [2] 24, p. 295; Mitscherlich, E. (1843). "Über die Gährung," *Berichte der Akademie der Wissenschaften Berlin*, pp. 35–41.

[38] [Anonymous] (1839). "Das enträthelte Geheimnis der geistigen Gährung," *Annalen der Chemie* 29, pp. 100–104.

In his 1836 annual report on the progress of chemistry, Jöns Jacob Berzelius (1779–1848) at the Karolinska Institutet in Stockholm, the most influential chemist of his time, wrote:

> . . . the conversion of sugar into carbonic and alcohol, as it occurs in the process of fermentation cannot be explained by a double decomposition-like chemical reaction between a sugar and so-called ferment, as we name the insoluble substance under the influence of which the fermentation takes place. This substance may be replaced by fibrin, coagulated plant protein, cheese and other materials, though the activities of these substances are at a lower level. However, of all the known reactions in the organic sphere, there is none to which the reaction bears a more striking resemblance that the decomposition of hydrogen peroxide under the influence of platinum, silver, or fibrin, and it would be natural to suppose a similar action in the case of the ferment.[39]

The "similar action" was the new force of catalysis:

> I do not consider this new force to be entirely independent of the electrochemical affinities of matter; I believe, on the contrary, that it is only a new manifestation of them, but so long as we cannot see their connection and mutual dependence, it will be more convenient to designate it by means of a separate name. I shall therefore term this force, catalytic force. I shall define catalysis as the decomposition of substances by this force, just as one defines analysis as the decomposition of substances by means of chemical affinity.[40]

Other prominent chemists did not welcome the word "catalysis" and for a time preferred the earlier term, "contact" substances, suggested by Mitscherlich to denote substances that accelerate chemical reactions without participating in them; among the reactions he listed as "decompositions or combinations by contact" were:

> . . . the breakdown of sugars to alcohol and carbonic acid, the oxidation of alcohol when it is converted to acetic acid, the reaction of urea and water to form carbonic acid and ammonia. As such, these substances undergo no change, but upon the addition of a small amount of ferment, which is the contact substance, and a definite temperature, these reactions take place at once.[41]

---

[39] Berzelius, J. J. (1836). "Einige Ideen über bei der Bildung organischer Verbindungen in der lebenden Natur wirksame, aber bisher nicht bemerkte Kraft," *Jahres-Berichte* 15, pp. 237–245 (240).

[40] *Ibid.*, p. 243.

[41] Mitscherlich, E. (1834). "Ueber die Aetherbildung," *Annalen der Physik* 31, pp. 273–282 (281).

Mitscherlich's views on fermentation were ridiculed by Justus Liebig (1803–1873), who advised him "not to burden chemists further with his old wives' gossip (Altweibergeschwätz), and to stop seizing upon the results of investigations he did not perform."[42]

Liebig has been considered by some historians to have been the greatest chemist of the nineteenth century.[43] In a lengthy paper published in 1839, perhaps in imitation of Berzelius, he wrote:

> I now wish to call attention of natural scientists to a hitherto unnoticed cause, through whose action are effected the phenomena of metamorphosis and decomposition generally denoted as decay, putrefaction, fermentation, and mouldering. This cause is ability of a substance in decomposition or combination, *i.e.* undergoing chemical reaction, to evoke the same reaction in another substance with which it is in contact, or to enable that substance to undergo the same changes that it undergoes itself.[44]

Accordingly, he considered yeast to be oxidized gluten in a state of putrefaction, and to communicate its reactivity to the decomposition of sugar. Liebig's ideas were widely disseminated in his popular writings, and appeared in textbooks of chemistry. For example, Charles Frédéric Gerhardt (1816–1856) defined a ferment as "any body that is in a state of decomposition and which by its contact with another substance promotes chemical changes in the latter."[45] The similarity of Liebig's definition to that of Georg Ernst Stahl in his *Zymotechnia fundamentalis* (1697) is striking.

The publicity given Liebig's theory may have stimulated young Hermann Helmholtz (1821–1894) in 1843 to study putrefaction and fermentation.[46] He showed that if a yeast suspension is separated from a sugar solution by means of a parchment membrane, fer-

---

[42] Liebig, J. (1842). "Mitscherlich und die Gährungstheorie," *Annalen der Chemie* 41, pp. 357–358.

[43] Brock, W. (1997). *Justus von Liebig. The Chemical Gatekeeper*. Cambridge University Press.

[44] Liebig, J. (1839). "Ueber die Erscheinungen der Gährung, Fäulnis und Verwesung und ihre Ursachen," *Annalen der Chemie* 30, pp. 250–287 (262).

[45] Gerhardt, C. F. (1856). *Traité de Chimie Organique*. Vol. IV, p. 541. Paris: Firmin-Didot.

[46] Helmholtz, H. (1843), "Ueber das Wesen der Fäulnis und Gährung," *Arch. Physiol.* 5, pp. 453–462. See McDonald, P. (2001). "Remarks on the context of Helmholtz's 'Ueber das Wesen der Fäulnis und Gährung'," *Science in Context* 14, pp. 493–498.

mentation only occurs in contact with the yeast, not on the other side of the membrane.

In parallel with the debate during 1830–1850 about the organismic theory of alcoholic fermentation, developments in the study of mammalian digestion provided greater clarity about the role of individual ferments. In 1833, Anselme Payen (1795–1871) and Jean François Persoz (1805–1865) reported that the addition of alcohol to an aqueous extract of germinating barley (malt) precipitated a flocculent material which, when dried and redissolved in water, could liquefy and convert it into sugar.[47] They called this material "diastase" (Gr. *diastasis*, separation). Before this report, in 1785 William Irvine (1743–1807) had shown that the addition of malt to a fermentation mixture increased the yield of sugar.[48] In 1815, Gottlieb Sigismund Constantin Kirchhoff (1764–1833) had found that an aqueous extract of dry malt could convert starch into sugar, and attributed the effect to some property of the gluten in the malt.[49] Also, in 1831, Erhard Friedrich Leuchs (1810–1837) demonstrated the conversion of starch to sugar by human saliva.[50] Subsequently, diastase activity was found in a variety of plants in the animal pancreas, renamed "amylase" (Lat., *amylum*, starch), and the term "diastase" was used in France to denote "soluble ferments."

The role of ferments in the digestion of food by animals had been the subject of speculation since Van Helmont, and mention was made above of the experimental studies by Reaumur and Spallanzani on the digestion of meat in the stomach of birds. The identification of a soluble gastric ferment able to convert foods such as egg albumin, the "fibrin" of meat, or coagulated blood, is rightly attributed to Johann Nepomuk Eberle (1795–1834), a physician in Würzburg.[51]

---

[47] Payen, A. and Persoz, J. F. (1833). "Mémoire sur la diastase etc.," *Annales de Chemie et de Physique* 53, pp. 73–92. An English translation is available in Boyde, T. R. C. (1980). *Foundation Stones of Biochemistry*, pp. 45–58. Hong Kong: Voile et Aviron.

[48] Irvine, W. (1805). *Essays, Chiefly on Chemical Subjects*, W. Irvine, Jr. (ed.). London: Mawman.

[49] Kirchhoff, G. S. C. (1815). "Ueber die Zuckerbildung beim Malzen des Getreides etc.," *Schweiggers Journal der Chemie und Physik* 14, pp. 389–398.

[50] Leuchs, E. F. (1831). "Wirkung des Speichels auf Stärke," *Annalen der Physik und Chemie* 22, p. 623.

[51] Eberle, J. N. (1834). *Physiologie der Verdauung nach Versuchen auf natürlichen und künstlichen Wege.* Wurzburg: Etlinger; Davenport, H. W. (1991). "Who was Johann Eberle?" *Gastroenterology International* 4, pp. 39–40.

He was stimulated to undertake this work by the extensive studies on animal digestion reported during the 1820s by Friedrich Tiedemann (1781–1861) and Leopold Gmelin (1788–1853), respectively professors of physiology and chemistry at Heidelberg.[52] Eberle discovered that one could prepare from dried gastric mucosa an acidified aqueous extract with a solvent power comparable to that exhibited by natural gastric juice. He concluded that "the mucosa does not merely contribute in a mechanical way to the chymification, but rather there exists between it and the foodstuff a special, unfortunately yet unknown, chemical relationship."[53]

Eberle probably had not learned before his death in December 1834 of the work of the American military surgeon William Beaumont (1785–1853), whose report on his studies on gastric digestion during 1825 to 1833 appeared in December 1833 (a German translation came out in 1834). Beaumont's experimental subject, Alexis St. Martin, had incurred a gunshot wound which created an opening (fistula) from the stomach to his left side, enabling Beaumont to examine samples of gastric juice after the ingestion of various nutrients. He concluded, from one of his final experiments that "Probably the gastric juice contains some principles inappreciable to the senses, or to chemical tests, besides the alkaline substances already discovered in it."[54] Like Tiedemann and Gmelin, Eberle and Beaumont confirmed the report of William Prout (1785–1850) that the chief acid of gastric juice is hydrochloric acid.[55]

Eberle's report attracted the attention of Johannes Müller (1801–1855), who had just become professor of anatomy and physiology in Berlin. During the winter of 1834/1835, he confirmed Eberle's findings on the action of the gastric mucosa on coagulated egg albumin, and then turned the problem over to Theodor Schwann.[56] The

---

[52] Mani, N. (1956). "Das Werk von Friedrich Tiedemann und Leopold Gmelin 'Das Verdauung nach Versuchen' und seine Bedeutung für die Entwicklung der Ernährungslehre in der ersten Hälfte des 19. Jahrhunderts," *Gesnerus* 13, pp. 190–214.

[53] Eberle (1834) (note 207), p. 160.

[54] Beaumont, W. (1833). *Experiments and Observations on Gastric Juice and the Physiology of Digestion*, p. 228. Plattsburg: Allen.

[55] Baron, J. H. (1979). "The discovery of gastric acid," *Gastroenterology* 76, pp. 1056–1064; Davenport, H. W. (1992), *A History of Gastric Secretion and Digestion.* New York: Oxford University Press.

[56] Müller, J. and Schwann, T. (1836). "Versuche über die künstliche Verdauung des geronnen Eiweisses," *Archiv für Anatomie und Physiologie*, pp. 66–89.

report by Schwann is frequently included among the great scientific papers of the nineteenth century.[57] He showed that a digestive fluid prepared by treating the mucosa membrane with hydrochloric acid and acetic acid effected digestion after filtration through linen cloth. He then filtered this fluid through paper and obtained an entirely clear yellowish filtrate with undiminished digestive power. This solubilization of the active principle was a considerable technical advance, and Schwann's repetition of the chemical tests applied by Tiedemann, Gmelin and Eberle to gastric juice and gastric mucosa strengthened the evidence for the existence of a separate chemical entity which he named *pepsin.* He also raised the question whether gastric digestion of albumin is comparable to the fermentation of sugar by yeast, and noted that in both processes a small amount of the active principle can convert a large amount of substrate.[58]

Schwann's pepsin paper received a mixed reception from his contemporaries. Berzelius wrote:

> The general conclusion that may be drawn at present from these experiments is that a very dilute solution which contains in addition to free hydrochloric acid a curious substance, Schwann's pepsin, simultaneously partly dissolves, partly extracts, and catalyzes the ingested nutrients. To obtain more precise and more certain results it is essential to isolate pepsin and to study its properties in the pure state.[59]

A short version of Schwann's paper, in which he referred to "catalytic or contact actions" appeared in the journal edited by Liebig, who appended the following footnote:

> Schwann's observations must lead to remarkable and interesting results, I do not wish however to leave unmentioned that they will be correctly understood when the substances taken up by hydrochloric acid have been prepared, and the changes in the albumin effected by its action have been studied by means of elementary analysis The name *pepsin* is provisionally only the representative of an idea, and before we decide to introduce the word *catalysis* into these investigations all means must be exhausted to solve the puzzle through analysis.[60]

---

[57] Schwann, T. (1836a). "Ueber das Wesen des Verdauungsprocesses," *Archiv für Anatomie und Physiologie*, pp. 90–138.

[58] *Ibid.*, p. 110.

[59] Berzelius, J. J. (1840). *Lehrbuch der Chemie.* Third edition, vol. 9, p. 215. Dresden: Arnold.

[60] Schwann, T. (1836b). "Ueber das Wesen des Verdauungsprocesses," *Annalen der Pharmacie* 20, pp. 28–34 (33–34).

In 1837, Liebig and his friend Friedrich Wöhler (1800–1882) added a third soluble ferment, which they named "emulsin." It was obtained from almonds, and catalyzed the hydrolysis of amygdalin to benzaldehyde, sugar, and hydrocyanic acid.[61] Emulsin was the first soluble ferment to be described as having an action on a well-defined crystalline compound whose composition was largely elucidated during the nineteenth century.

During 1820–1850 there was a lively Franco-German competition in the study of digestion. The French adversaries of Tiedemann and Gmelin were François Leuret (1797–1851) and Jean Louis Lassaigne (1800–1851), who claimed that the chief gastric acid is lactic acid. In 1840, Jean Baptiste Deschamps (1804–1866) claimed that the agent he named "chymosine" is the specific gastric agent in the chymification process.[62] In 1845, the physiologist Claude Bernard (1813–1878) and his chemical associate Charles Louis Barreswil (1817–1870) supported the claim for lactic acid. They also inferred from the fact the pancreatic juice, gastric juice, and saliva, converted starch into sugar when the mixture was alkaline, and digested meat when it was acidic, that

> there exists one active principle in digestion, which is common to them, and that it is only the nature of the chemical reaction which causes the physiological role of each of these liquids to differ, and which determines their digestive aptitude for one or another alimentary principle.[63]

This inauspicious beginning of Bernard's scientific career was of course overshadowed by his later brilliant achievements. The idea of a unitary ferment whose specificity depends on acidity or alkalinity of the medium was contradicted by Louis Mialhe (1807–1886):

> Each of the ferments has an action appropriate to itself. One of them, salivary diastase, liquefies starch transforms it into dextrin an glucose in less than a minute, another, pepsin, which possesses no saccharify-

[61] Liebig, J. and Wöhler, F. (1837). "Über die Bildung des Bittermandelöls," *Annalen der Pharmacie* 22, pp. 1–24.

[62] Deschamps, J. B. (1840). "De la présure," *Journal de Pharmacie* 26, pp. 412–420.

[63] Bernard, C. and Barreswil, C. L. (1845). "Recherches expérimentales sur les phénomènes chimiques de la digestion," *Compt. Rend.* 21, pp. 88–89. See Sernka, T. J. (1979). "Claude Bernard and the nature of gastric acid," *Perspectives in Biology and Medicine* 22, pp. 523–530.

> ing action on starch, coagulates milk, fibrin, and gluten; then dissolves the coagulum, and subjects it to a very particular molecular transformation. On the other hand, diastase exerts no action whatever on albuminoid fluids ... It is not possible to agree with Liebig, Bernard, Barreswil, and others, that the ferments are instantly produced and destroyed as soon as the need for them is felt, or that these ferments are one and the same principle which exhibits different qualities depending on the medium in which it is placed, and depending on substance to which it is exposed. For us, these materials are special and distinct, each one conserving its nature, its particular role, and its complete independence ... Up to the present, we know only two, diastase and pepsin, in animals, but there certainly exist others which also participate in the maintenance of life.[64]

In 1848, Bernard identified in pancreatic juice a ferment which emulsified and saponified fats; it was later named a "lipase."[65]

It would appear that by the 1850s, the issue was clearly drawn between the upholders of the organismic theory of vinous fermentation and those insisting on the individuality of animal ferments. Some ambiguity was introduced into the debate, however, by the use of the term "vital force" by Berzelius and Liebig.[66] A different approach was offered in 1858 by Moritz Traube (1828–1894), a professional chemist and manager of the family brewery. He noted that

> Even if all fermentations depended on the presence of infusoria or fungi, a healthy science would not block the road to further research by means of such a hypothesis; it would conclude from these facts that there are present in these microscopic organisms certain chemical substances which elicit the phenomena of decomposition. It would attempt to isolate these substances, and if it could not isolate them without change in their properties, it would only conclude that the separation methods had exerted a deleterious effect on these substances.[67]

Indeed, in 1846, Lüdersdorff reported that he had ground yeast on a glass plate until no globules could be seen under a microscope,

---

[64] Mialhe, B. (1856). *Chimie Appliquée à la Physiologie et à la Thérapeutique*, pp. 35–36. Paris: Masson.

[65] de Romo, A. C. (1989). "Tallow and the time capsule: Claude Bernard's discovery of the pancreatic digestion of fat," *History and Philosophy of the Life Sciences* 11, pp. 251–274.

[66] Jørgensen, B. S. (1964). "Berzelius und die Lebenskraft," *Centaurus* 10, pp. 258–281; Lipman, T. O. (1967). "Vitalism and reductionism in Liebig's physiological thought," *Isis* 58, pp. 167–185; Hall, M. D. V. (1980). "The role of force or power in Liebig's physiological chemistry," *Medical History* 24, pp. 20–59.

[67] Traube, M. (1858). *Theorie der Fermentwirkungen*, pp. 7–8. Berlin: Dummler.

and that the resulting material failed to ferment glucose. In the following year, Carl Schmidt (1822–1894) repeated these experiments with longer trituration and explained the negative result as due to the destruction of the ferment; during the course of this work Schmidt anticipated Pasteur's finding of succinic acid as a regular product in alcoholic fermentation.[68]

In 1857, Louis Pasteur (1822–1895) introduced his first report to the *Académie des Sciences* on lactic acid fermentation as follows:

> I was led to occupy myself with fermentation as a consequence of my research on the properties of the amyl alcohols and the very remarkable crystallographic particulars of their derivatives. I will have the honor later of presenting to the *Académie* observations which offer an unexpected link between the phenomena of fermentation and the property of molecular dissymmetry characteristic of natural organic substance.[69]

After his famous work on the molecular dissymmetry of the tartrates, Pasteur studied the optical activity of asparagine, aspartic acid, malic acid, and amyl alcohol. The results led him to the conviction that, if an organic substance possesses optical activity, it must have been formed by a physiological process. In discussing the difference between organic substances obtained from biological sources and those made in the laboratory, he stated: "The artificial products do not have any molecular dissymmetry; and I could not indicate the existence of a more profound separation between the products born under the influence of life, and all the others."[70]

Pasteur was an extremely skillful experimenter, and endowed with great ability to attack controversial problems by selecting for criticism the weak points in a theory he wished to disprove.[71] These equalities were evident in his first paper on fermentation. Much was known about the conversion of sugar to lactic acid. Lactic acid had

---

[68] Schmidt, C. (1862). "Zur Geschichte der Gährung," *Annalen der Chemie* 126, pp. 126–128.

[69] Pasteur, L. (1922). *Oeuvres de Pasteur*. Vol. 2, p. 14. Paris: Masson.

[70] Pasteur, L. (1861). "Recherches sur la dissymmétrie moléculaire des produits organiques naturels" in: *Société de Chimique de Paris, Leçons de Chimieprofessées en 1860*, pp. 1–48 (33).

[71] Duclaux, E. (1896). *Pasteur: Histoire d'un Esprit*. Sceaux: Charaire; Dubos, R. (1950). *Louis Pasteur: Free Lance of Science*. Boston: Little Brown; Geison, G. L. (1995). *The Private Science of Louis Pasteur*. Princeton University Press; Paul, H. W. (1996) (note 183), pp. 155–193.

been isolated by Scheele from soured milk in 1780.[72] Several French chemists had shown during the 1840s that the addition of chalk to a fermentation mixture markedly increases the yield of lactic acid, but in contrast to vinous fermentation no one had yet claimed to have identified a microbial agent for lactic fermentation, and a labile albuminoid material was considered to be the catalytic agent.[73] Pasteur therefore announced that

> I intend to establish in the first part of this work that, just as there is an alcoholic ferment, the yeast of beer, which is found everywhere where sugar is decomposed to alcohol and carbonic acid, so also there is a particular ferment, a lactic yeast, always present when sugar becomes lactic acid, and if all labile nitrogenous material can transform sugar into this acid, it is because it is a suitable nutrient for the development of this ferment.[74]

The paper on the lactic ferment, seen under the microscope as globules smaller than beer yeast, was followed during 1857–1859 by a rapid succession of preliminary notes on alcoholic fermentation. Liebig's theory of the generation of yeast by the oxidation of nitrogenous material was disproved by producing yeast in an aqueous medium containing only sugar, ammonium tartrate, and a mineral phosphate. Pasteur showed that, contrary to earlier views, ammonia is not a normal product of the fermentation of sugar by yeast, but rather that the ammonia is transformed into complex albuminoid material, which enters the structure of the yeast. He also denied that there is a chemical equation, such as that of Lavoisier and Gay-Lussac, for the fermentation of sugar to alcohol and carbonic acid, since he also found succinic acid and glycerine to be normal products.

In 1860 Pasteur presented an extended and detailed report of his studies on alcoholic fermentation. He concluded from his analyses of the products that

> The variations in the proportions of succinic acid, or glycerine, and consequently of the other products of fermentation, should not be surprising in a phenomenon in which the conditions contributed by the

[72] Scheele, C. W. (1793). *Sämmtliche Physische und Chemische Werke.* Vol. 2, pp. 249–265. Berlin: Martin Sändig.

[73] Pelouze, J. and Gélis, A. (1843). "Mémoire sur l'acide butynque," *Comptes Rendus* 16, pp. 1262–1271.

[74] Pasteur, L. (1922), p. 6.

ferment seem of necessity to be so changeable. What has surprised me, on the contrary, is the usual constancy of the results. The various analyses in this article provide us enough proof of this.

I am therefore much inclined to see in the act of alcoholic fermentation a phenomenon which is simple, unique, but very complex, as it can be for a phenomenon correlative with life, giving rise to multiple products, all of which are necessary. The globules of yeast, true living cells, may be considered to have as the physiological function correlative with their life the transformation of sugar, somewhat like the cells of the mammary gland transform the elements of the blood into the various constituents of milk, correlatively with their life and the changes in their tissues.

My present and most fixed opinion regarding the nature of alcoholic fermentation is this: The chemical act of fermentation is essentially a phenomenon correlative with a vital act, beginning and ending with the latter. I believe that there is never any alcoholic fermentation without there being simultaneously the organization, development, multiplication of the globules, or the pursued, continued life of globules that are already formed. The totality of the results in this article seems to me to be in complete opposition to the opinions of MM. Liebig and Berzelius.

I profess the same views on the subject of lactic fermentation, butyric fermentation, the fermentation of tartaric acid, and many other fermentations properly designated as such (*fermentations proprement dites*) that I shall study successively.

Now, what does the chemical act of the cleavage of sugar represent for me, and what is its intimate cause? I confess that I am completely ignorant of it.

Will you say that the yeast nourishes itself with sugar so as to excrete it in the form of alcohol and carbonic acid? Will you say on the contrary that the yeast produces, during its development, a substance such as pepsin, which acts on the sugar and disappears when that is exhausted, since one finds no such substance in the liquids? I have no reply on the subject of these hypotheses. I do not accept them or reject them, and wish to constrain myself always not to go beyond the facts. And the facts only tell me that all the fermentations properly designated as such are correlative with physiological phenomena.[75]

Pasteur then described the ferment which transformed sugar into butyric acid as an organism that is killed by free oxygen, and he defined fermentation as life without air (*vie sans air*). He also observed that oxygen inhibits the conversion of sugar to alcohol, and gives more cells per unit of sugar consumed. This phenomenon, linking fermen-

[75] Pasteur, L. (1922), pp. 51–126 (76–77).

tation to respiration, has been termed the "Pasteur effect" and has puzzled biochemical investigators for decades.[76]

In 1860 Pasteur encountered an adversary in the chemist Marcelin Berthelot (1827–1907), whose objective was: ". . . in every fermentation, one must try to reproduce the same phenomena by chemical methods and to interpret them by exclusively mechanical considerations. To banish life from all explanations relative to organic chemistry, that is the aim of our studies."[77] Berthelot chose the discovery by Augustin Pierre Dubrunfaut (1797–1881) of the conversion of cane sugar (sucrose) by dilute acid into "invert sugar" (glucose and fructose).[78] This process can be followed by polarimetry, since there is an "inversion" in the optical rotation of the reaction mixture because the dextrorotation of sucrose and glucose is much less than the levorotation of fructose. Berthelot showed that it is possible to extract from yeast a water soluble *ferment glucosique* which effects this optical inversion.[79] Pierre Béchamp (1816–1908) found this ferment in many plants, and called it "zymase"; it was later renamed "invertase." Pasteur had suggested that the inversion might be effected by the succinic acid which appeared in alcoholic fermentation, but Berthelot disproved this idea. His paper ends with the statement that

> If a deeper study leads to the extension of the view which I propose, and to its application with certainty to the insoluble ferments, all fermentations would be brought under one the same general concept, and they could be definitely assimilated to effects of acids provokes by contact, and of truly chemical reagents.[80]

Pasteur's prompt rejoinder was that Berthelot

> . . . here calls *ferment* substances soluble in water, and able to invert sugar. Now everyone knows that there are very many (*une foule de*) substances that enjoy this property, for example all the acids. As for me,

[76] Krebs, H. A. (1972). "The Pasteur effect and the relations between respiration and fermentation," *Essays in Biochemistry* 8, pp. 1–34.
[77] Berthelot, M. (1860). *Chimie Organique Fondée sur la Synthèse*. Vol. 2, p. 656. Paris: Mallet-Bachelier.
[78] Dubrunfaut, A. P. (1846). "Note sur quelques phénomènes rotatoires et sur quelques propriétes des sucres," *Annales de Chimie* [3] 18, pp. 99–108.
[79] Berthelot, M. (1860). "Sur la fermentation glucosique du sucre de canne," *Comptes Rendus* 50, pp. 980–984.
[80] *Ibid.*, p. 984.

> when it is a question of cane sugar and beer yeast, I only term ferment that which causes the fermentation of sugar, that is to say which produces alcohol, carbonic acid, etc. As to inversion, I have not occupied myself with it. In regard to its cause, I only proposed a doubt in passing, in a note in a memoir in which I summarized three years of study on alcoholic fermentation.[81]

Another adversary of the 1860s was Béchamp, who claimed priority in the discovery of airborne "molds" (*moisissures*) as the ferments in the alcoholic fermentation of sugar.[82] In the debates with his critics, Pasteur chose his words in a manner best calculated to accord with his preconceived ideas. The use of the term "fermentations proprement dites" narrows the discourse to fermentations caused by living organisms, and is a virtual tautology.[83]

Pasteur's next opponent was the famous Justus von Liebig, who published in 1870 a three part paper on alcoholic fermentation, acetic fermentation, and the source of muscular power. Liebig began by reiterating his theory:

> I assumed that the breakdown of the fermentable substance to simpler compounds must be explained by a process of cleavage residing in the ferment... The rearrangement of the sugar molecule is therefore a consequence of the decomposition or rearrangement of one or several constituents of the ferment, and occurs only when they are in contact.[84]

Liebig also questioned an experiment in which Pasteur used a small amount of yeast. Pasteur's reply was brief and scornful; according to his biographer, Liebig was very upset.[85]

During the 1870s, there were also disputes with Edmond Fremy (1814–1894) and Oscar Brefeld (1839–1924), who objected to Pasteur's

---

[81] Pasteur, L. (1922), pp. 127–128. See Geison, G. L. (1981). "Pasteur on vital versus chemical ferments: A previously unpublished paper on the inversion of sugar," *Isis* 72, pp. 425–445.

[82] Béchamp. A. (1855). "De l'influence que l'eau pur ou chargée de sels exerce à froid sur le sucre de canne," *Comptes Rendus* 40, pp. 44–47. See Nonclercq, M. (1982). *Antoine Béchamp*. Paris: Maloine; Manchester, K. L. (2001). "Antoine Béchamp: père de la biologie. Oui ou non?" *Endeavour* 25, pp. 68–73.

[83] Temple, D. (1986). "Pasteur's theory of fermentation: a virtual tautology?" *Studies in the History and Philosophy of Science* 17, pp. 487–503.

[84] Liebig, J. (1870). "Über die Gährung und die Quelle der Muskelkraft," *Annalen der Chemie und Pharmacie* 153, pp. 1–47, 137–229 (1).

[85] Volhard, J. (1909). *Justus von Liebig*. Vol. 2, pp. 88–103. Leipzig: Barth.

restriction of the *ferments proprement dites* to airborne microorganisms.[86] Of particular importance were the studies of Georges Vital Lechartier (1837–1909) and Félix Bellamy, which were confirmed by Pasteur, on the alcoholic fermentation in a variety of plants and in the absence of microorganisms.[87] The most dramatic encounter was the reappearance in 1878 of Berthelot as Pasteur's critic, by the publication in the 20 July issue of the *Revue Scientifique* of the experimental notes of the great physiologist Claude Bernard, who died on the previous 10 February. During the fall of 1877 Bernard had studied the fermentation of the juice of rotting fruit, and had apparently believed that he had shown fermentation to occur without the participation of living cells and the conversion of sugar into alcohol could be effected by agents separable from living yeast. In his last book, he reiterated his conviction that a distinction must be made between chemical synthesis as a phenomenon of life, and chemical degradation as a process independent of life:

> In my view, there are two orders of phenomena in the living organism: 1) The phenomena of *vital creation* or *organizing synthesis*; 2) The phenomena of death or *organic destruction*... The first of these two orders of phenomena is alone without direct analogues; it is specific (*particulier, spécial*) to the living being: This evolving synthesis is what is truly vital—I shall recall in this connection the formula I expressed a long time ago: "*Life is creation.*" The second, namely vital destruction, is on the contrary of a physico-chemical order, most often the result of a combustion, of a fermentation, of a putrefaction, in a word of an action comparable to a large number of chemical decompositions or cleavages. These are the true phenomena of *death*, as applied to the organized being. And, it is worthy of note that we are here the victims of a habitual illusion, and when we wish to designate the phenomena of *life*, we in fact indicate the phenomena of *death*.[88]

It is clear from this excerpt that Bernard could not accept Pasteur's designation as a phenomenon correlative with life. It is also clear

---

[86] Fremy, E. (1875). *Sur la Génération des Ferments*. Paris: Masson; Höxtermann, E. (1997). "Oscar Brefeld (1839–1925) and the complementary perspective of chemistry and botany toward alcoholic fermentation in the 1870s" in: *Biology Integrating Scientific Disciplines*, B. Hoppe (ed.), pp. 174–188. Munich: Institut für die Geschichte der Wissenschaft.

[87] Lechartier, G. V. and Bellamy, F. (1875). "De la fermentation des fruits," *Comptes Rendus* 81, pp. 1129–1132. See Green, J. R. (1901). *The Soluble Ferments and Fermentations*. 2nd ed, pp. 348–349. Cambridge University Press.

[88] Bernard, C. (1878–1879). *Leçons sur les Phénomènes de la Vie Communs aux Animaux aux Végétaux*. Vol. 1, pp. 39–40. Paris: Baillière.

that Pasteur was talking about the life of microorganisms, while Bernard was writing about higher animals and plants. Pasteur's first reply to the notes published by Berthelot came only two days later, on 22 July. There then followed an acerbic but inconclusive debate at the *Académie des Sciences*, during which Pasteur stated:

> I must add finally that it is always an enigma to me that one could believe that I would be disturbed by the discovery of soluble ferments in the fermentations properly designated as such, or by the formation of alcohol from sugar, independently of living cells. Certainly, I confess it without hesitation, and if one wishes, I am ready to explain myself on this point at greater length, I do not see either the necessity for the existence of these ferments or the utility of their function in this kind of fermentation.[89]

Another kind of fermentation considered by Pasteur to belong to those "proprement dites" was formation in urine of ammonia from urea. In 1863 he wrote: "I am led to believe that this production constitutes an organized ferment and that there is never any transformation of urea into ammonium carbonate without the presence and development of this small plant."[90] In 1876, Frédéric Alphonse Musculus (1829–1888) reported that he had isolated from ammoniacal human urine a "soluble ferment" which readily converted urea into ammonium carbonate.[91] This finding was confirmed by Pasteur, who wrote:

> The result that we have just announced at the *Académie* was not and could not have been foreseen. It is the first example of an autonomous organized ferment whose function merges with one of its unorganized products. It is also a new example of a *diastase* produced during life and able to modify a substance by the fixation of water, in the same manner as all the other *diastases*.[92]

Pasteur also modified his theory of "vie sans air" to include oxygen in the initial stage of alcoholic fermentation. The noted Franco-Croat historian of science Mirko Drazen Grmek (1924–2001) wrote that Pasteur

---

[89] Pasteur, L. (1922), p. 592.

[90] Pasteur, L. (1922), p. 249.

[91] Musculus, F. A. (1876). "Sur le ferment de l'urée," *Comptes Rendus* 82, pp. 333–336.

[92] Pasteur, L. (1922). *Oeuvres de Pasteur.* Vol. 6, p. 85. Paris: Masson.

> experimented to verify his intuitions, in order to provide irrefutable proof of his fundamental ideas. If an experiment did not confirm his initial hypothesis, he did not do any more experiments in that direction. Pasteur's very particular genius consisted in the historical fact that he nearly always was at once correct. He thought much before experimenting, he presented precise questions before nature.[93]

In 1863 the Emperor (Napoleon III) asked him to look into the diseases of wine, and the method he proposed (heating at about 55°) became known as "pasteurization." One of Pasteur's best students, Ulysse Gayon (1845–1929) became head of a leading laboratory of oenology at Bordeaux.[94] At the newly-established laboratory of the Carlsberg brewery in Copenhagen, Emil Christian Hansen (1842–1909) was an admirer of Pasteur, and in 1883 introduced the use of a single yeast cell to generate "pure yeast" (named Carlsberg bottom yeast I). English brewers, largely indifferent to Pasteur's work, became interested in Hansen's approach, but adopted it only partially.[95]

To sum up this sketch of Pasteur's role in the fermentation story, I quote the scholar of oenology Harry Paul:

> It seems difficult for the French to subject Pasteur to critical scrutiny. He has icon status. Not that it is difficult for his admirers to admit that some of his ideas have been shown to be incomplete or even wrong. Two obvious cases are the Pasteurian view of fermentation as an exclusively biological or vital process, excluding the enzymes produced by the yeast, and his simplistic hypothesis that the aging of wine is essentially the combining of oxygen with the wine.[96]

I now return to Moritz Traube, whose quoted statement in 1858 did not include Pasteur's name. In 1878, that statement appears in the following revised form:

> 1) The ferments are not, as Liebig assumed, substances in a state of decomposition, and which can transmit to ordinarily inert substances their chemical action, but are chemical substances related to the

---

[93] Grmek, M. D. (1991). "Louis Pasteur, Claude Bemard et la méthode expérimentale" in: *L'Institut Pasteur*, M. Morange (ed.), pp. 21–44 (29). Paris: La Découverte.
[94] Paul, H. W. (1996), pp. 275–287.
[95] Klöcker, A. (1976). "Emil Christian Hansen" in: *The Carlsberg Laboratory 1876–1976*, pp. 168–189. Copenhagen: Rhodos; Teich, M. (1983). "Fermentation theory and practice: the beginnings of pure yeast cultivation and English brewing, 1883–1913," *History of Technology* 8, N. Smith (ed.), pp. 117–133.
[96] Paul, H. W. (1996), p. 191.

> albuminoid bodies which, not yet accessible in pure form, have like all other substances a definite chemical composition and evoke changes in other substances through definite chemical affinities. 2) Schwann's hypothesis (later adopted by Pasteur), according to which fermentations are to be regarded as expressions of the vital forces of lower organisms is unsatisfactory . . . The reverse of Schwann's hypothesis is correct: Ferments are the causes of the most important vital-chemical processes, and not only in lower organisms, but in higher organisms as well.[97]

In a paper confirming Pasteur's experimental evidence for anaerobic fermentation, Traube also restated his 1858 theory of intermolecular oxygen transfer:

> I have shown by means of numerous examples that just as there are substances which like platinum, nitric oxide (in sulfuric acid manufacture), indigo sulfuric acid, etc., can transfer free oxygen to other substances and effect their oxidation (oxygen carriers, oxidation ferments), there are also substances that can transfer bound oxygen, that is they can effect reduction of one part and oxidation of the other. If we imagine the sugar molecule to be composed of 2 atomic groupings, a reducible A and an oxidizable B, then the cleavage by the yeast ferment is effected in such a manner that it extracts oxygen from group A (the deoxidized product is alcohol) in order to transfer it to group B, which is thereby burned to carbonic acid.[98]

A somewhat similar theory of fermentation was proposed in 1874 by Felix Hoppe-Seyler (1825–1895), but which assumed that "all reductions occurring in putrefying fluids are secondary processes elicited by nascent hydrogen", and that when oxygen is present, "instead of the reduction, there appears during the putrefaction oxidation, which can have its cause in nothing but the cleavage of the oxygen molecule by the nascent hydrogen . . . whereby the oxygen is converted to an activated state and can then act as a powerful oxidizing agent."[99] Hoppe-Seyler assumed that the oxygen molecule

[97] Traube, M. (1878). "Die chemische Theorie der Fermentwirkungen und der Chemismus der Respiration," *Berichte der deutschen chemischen Gesellschaft* 11, 1984–1992 (1984). See Sourkes, T. L. (1955). "Moritz Traube (1826–1894). His contribution to biochemistry," *Journal of the History of Medicine* 10, pp. 379–391.

[98] Traube, M. (1874). "Ueber das Verhalten der Alkoholhefe in sauerstoffgasfreien Medien," *Berichte der deutschen chemischen Gesellschaft* 7, pp. 872–887 (884).

[99] Hoppe-Seyler, F. (1876). "Ueber die Processe der Gährungen und ihre Beziehung zum Leben des Organismus," *Pflügers Archiv* 12, pp. 1–17 (15–16).

(O_2) had to be cleaved and that the formation of hydrogen peroxide (H_2O_2) during the oxidation of organic substances was a consequence of the addition of activated atomic oxygen to water. Traube disproved this idea by showing that in *Autoxidation* (his term for oxidation by oxygen gas) there is no cleavage of O_2 to activated atomic oxygen, but rather the addition of molecular oxygen to the organic molecule to form what he called a "holoxide." The initial reception was mixed, partly because Traube did not occupy an important professorship, but Traube's theory received impressive experimental support from subsequent chemical studies.[100]

In 1878, the physiologist Friedrich Wilhelm (Willy) Kühne (1837–1900) introduced the word "enzyme" into the discourse about fermentation. He suggested that the designations of organized and unorganized ferments

> have not gained wide acceptance, in that on the one hand it was explained that chemical bodies, like ptyalin, pepsin, etc., could not be called ferments, since the name was already assigned to yeast cells and other organisms (Brücke), while on the other hand it was said that yeast cells could not be called ferment, because then all organisms, including man, would have to be so designated (Hoppe-Seyler). Without wishing to inquire further why the name has generated so much excitement from opposing sides, I have taken the liberty, because of this contradiction, of giving the name enzymes to some of the better substances, called by many "unorganized ferments." This not intended to imply any particular hypothesis, but it merely states that in zyme (yeast) something occurs that exerts this or that activity, which is considered to belong to the class called fermentative. The name is not, however, intended to be limited to the invertin of yeast, but it is intended to imply that more complex organisms, from the enzymes pepsin, trypsin, etc, can be obtained, are not so fundamentally different from the unicellular organisms as Hoppe-Seyler, for example, appears to think.[101]

The words "enzyme" and "azyme" had been used during the disputes in the early Christian church over the question whether leavened or unleavened bread should be used for the Eucharist. Hoppe-Seyler wrote: "The new word may be added to the large number of new names that Kühne has proposed . . . for substances that are totally

[100] Milas, N. A. (1932). "Auto-oxidation," *Chemical Reviews* 10, pp. 295–364.
[101] Kühne, W. (1878). "Erfahrungen und Bemerkungen über Enzyme und Fermente," *Untersuchungen aus dem physiologischen Institut Heidelberg* 1, pp. 291–324 (293).

unknown."[102] Kühne had just given the name "trypsin" to the long known protein-cleaving enzyme in pancreatic juice. The word "enzyme" was adopted fairly readily in England and Germany, but "diastase" was retained in France. The adoption of "enzyme" by French scientists was accompanied by a gradual shift during the twentieth century from its original feminine gender to the masculine;[103] according to *Le Monde* of 3 June 1970, the *Académie Française* had ruled, on 3 February 1970, in favor of the feminine.

Traube's theory of fermentation elicited a negative judgement from the noted botanist Carl Nägeli (1817–1891):

> The agent of fermentation is inseparable from the substance of the living cell, i.e., it is linked to the plasma [Nägeli's word for protoplasm]. Fermentation occurs only in immediate contact with plasma in so far as its molecular action extends. If the organism wishes to exert an effect on chemical processes in places or at distances where the molecular forces of living matter are without power, it excretes ferments. The latter are especially active in the cavities of the animal body, in the water in which molds live, and in the plasma-poor cells of plants. It is even doubtful whether the organism ever makes ferments that are to function within the plasma; since here it does not need them, because it has available to it in the molecular forces of living matter much more energetic means for chemical action.[104]

Nägeli defined "plasma" as a semi-liquid mucilaginous content of the plant cell, which consists of variable amounts of insoluble and soluble albuminates, and defined fermentation as "the transmission of the state of motion of molecules, atomic groups, and atoms of the various compounds in the living plasma (which remain chemically unchanged) to the fermented material whereby the steady state in its molecules so disturbed are brought to decomposition."[105]

Another noted botanist, the younger Johannes Reinke (1849–1931), endorsed Traube's theory,[106] possibly because most of the known natural substances (guiac, indigo) thought to be intracellular oxygen car-

---

[102] Hoppe-Seyler, F. (1878). "Ueber Gährungsprocesse," *Zeitschrift für physiologische Chemie* 2, pp. 1–28 (3–4).

[103] Plantefol, L. (1968). "Le genre du mot enzyme," *Comptes Rendus* 266, pp. 41–46.

[104] Nägeli, C. (1878). "Theorie der Gärung," *Abhandlungen der königlichen Akademie der Wissenschaften* 13 (2), pp. 77–205 (86–87).

[105] *Ibid.*, p. 100.

[106] Reinke, J. (1883). "Die Autoxydation in der lebenden Pflanzenzelle," *Botanische Zeitung* 41, 65–76, pp. 89–103.

riers were derived from plants. Moreover, in 1883 Hikorokuro Yoshida (1859–1928) found the first oxidative enzyme, which promoted the darkening and hardening of the latex of the Japanese lacquer tree. This line of work was taken up by Gabriel Bertrand (1867–1962) at the Pasteur Institute, and by 1897 he had shown that there are numerous enzymes, which he named "oxidases," able to catalyze the activation of molecular oxygen and to exhibit specificity with respect to the substance undergoing oxidation.[107]

To the increase in the number and variety of individual ferments before 1900 must be added the parallel increase in the number and variety of microbial fermentations brought to light by the work of men such as Sergei Nikolaevich Vinogradsky (1856–1953) and Martinus Willem Beijerinck (1851–1931) through their use of enrichment cultures.[108] Among those who have been largely forgotten is Frédéric Diénert (1874–1948), who described in 1900 the adaptation of yeast to the fermentation of galactose.[109]

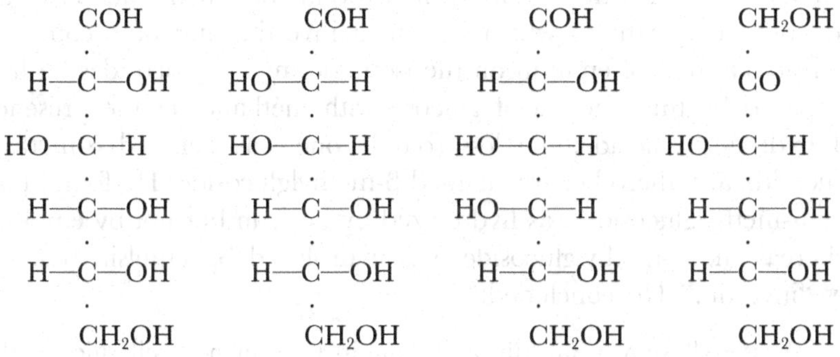

COH	COH	COH	CH_2OH
H—C—OH	HO—C—H	H—C—OH	CO
HO—C—H	HO—C—H	HO—C—H	HO—C—H
H—C—OH	H—C—OH	HO—C—H	H—C—OH
H—C—OH	H—C—OH	H—C—OH	H—C—OH
CH_2OH	CH_2OH	CH_2OH	CH_2OH
d—Glucose	*d*—Mannose	*d*—Galactose	*d*—Fructose

[107] Gaudillière, J. P. (1991). "Catalyse enzymatique et oxydations cellulaires. L'oeuvre de Gabriel Bertrand et son heritage" in: *L'Institut Pasteur*, M. Morange (ed.), pp. 118–136. Paris: La Découverte.

[108] Gutina, V. (1976). "Sergey Nikolaevich Vinogradsky," *Dictionary of Scientific Biography* 14, pp. 36–38. New York: Scribners; Hughes, S. S. (1978). "Martinus Willem Beijerinck," *ibid.*, 15, pp. 13–15.

[109] Diénert, F. V. (1900). *Sur la fermentation de galactose et sur l'accoutumance des levures à ce sucre*. Sceaux: Charaire.

Nägeli's approach to the fermentation problem was different from that of Emil Fischer (1852–1919), the most distinguished organic chemist of his generation. After ten years of synthetic work in his laboratory he had established the constitution and stereochemistry of the principal known monosaccharides, including glucose, fructose, marmose, and galactose (and their derivatives). In 1894 he and his biochemical colleague Hans Thierfelder (1858–1930) reported their findings on action of 12 different pure strains of brewers yeast, most of them the gift of Hansen in Copenhagen. They concluded that

> Among the agents used by the living cell, the principal role is played by the various albuminoid substances. They are optically active, and since they are synthesized from the carbohydrates of plants, one may well assume that the geometrical structure, as regards their asymmetry, is fairly similar to that of the natural hexoses. On the basis of this assumption, it would not be difficult to understand that the yeast cells, with their constructed agent, can only attack and ferment those kinds of sugars whose geometry is not too different from that of grape sugar.[110]

Later in 1894, Fischer reported his experiments on the action of an aqueous yeast extract (which he called invertin) and of a commercial preparation of emulsin on the isomeric methyl glucosides he had prepared by the reaction of glucose with methanol in the presence of hydrogen chloride; the less soluble one was denoted α-methylglucoside and the other was named β-methylglucoside. He found that the α-methylglucoside was hydrolyzed by invertin but not by emulsin, whereas the β-methylglucoside was hydrolyzed by emulsin but not by "invertin." He concluded:

> As is well known, invertin and emulsin have many similarities to the proteins and undoubtedly also possess an asymmetrically constructed structure. Their restricted action on the glucosides may therefore be explained on the basis of the assumption that only with a similar geometrical structure can the molecules approach each other closely, and thus initiate the chemical reaction. To use a picture, I would say that the enzyme and the glucoside must fit each other like a lock and key, in order to effect a chemical action on each other . . . The finding that the activity of enzymes is limited by molecular geometry to so marked

[110] Fischer, E. and H. Thierfelder (1894). "Verhaltung der verschieden Zucker gegen reine Hefen," *Berichte der deutschen chemischen Gesellschaft* 27, pp. 2031–2037 (2037).

a degree should be of some use for physiological research. Even more important for such research seems to me to be the demonstration that the difference frequently assumed in the past between the chemical activity of living cells and of chemical reagents, in regard to molecular asymmetry, is nonexistent.[111]

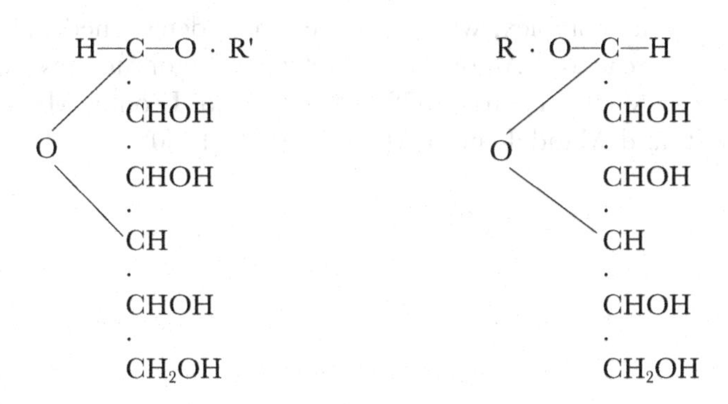

α- and β-methylglucosides

This theme was developed more fully in a famous 1898 review article.[112] Despite the uncertainty of the "invertin" in Fischer's experiment, the "lock and key" analogy gained immortality, and the concept of enzyme-substrate interaction became part of the modern physical-chemical theory of reaction kinetics and catalysis developed by Jacobus Henricus van 't Hoff (1852–1911), Svante Arrhenius (1859–1927), and Wilhelm Ostwald (1853–1932). For example, Van 't Hoff, who established the criteria for the equilibrium in chemical reactions, noted in 1898 that

> It follows from theoretical considerations that, in case a ferment is not changed during its action, a state of equilibrium and not a complete conversion must be attained. Hence it also follows that the reaction must be able to proceed in the opposite direction as well. One can justifiably ask whether (using equilibrium theory) formation of sugar from carbon dioxide and alcohol takes place under the influence of

[111] Fischer, E. (1894). "Einfluss der Konfiguration auf die Wirkung der Enzyme," *Berichte der deutschen chemischen Gesellschaft* 27, pp. 2985–2993 (2992–2993).

[112] Fischer, E. (1898). "Bedeutung der Stereochemie für die Physiologie," *Zeitschrift für physiologische Chemie* 26, pp. 60–87.

> zymase if one exceeds the final pressure of carbon dioxide, and whether trypsin may not under certain conditions given by equilibrium theory be able to form protein from the cleavage products it itself forms.[113]

Emil Fischer's approach also influenced the development of the theory of enzyme kinetics in terms of the intermediate formation of an enzyme-substrate complex, whose "active mass" determined the rate of the overall reaction. Among the chief contributors to this development were Victor Henri (1872–1940),[114] and Leonor Michaelis (1875–1949) and Maud Lenora Menten (1879–1960).[115]

---

[113] Hoff, J. H. van't (1898). "Über die zunehmende Bedeutung der anorganischen Chemie," *Zeitschrift für anorganische Chemie* 18, pp. 1–13 (12–13).

[114] Henri, V. (1903). *Lois Générales de L'Action des Diastases*. Paris: Hermann.

[115] Michaelis, L. and M. L. Menten (1913). "Zur Kinetik der Invertinwirkung," *Biochemische Zeitschrift* 49, pp. 333–336.

CHAPTER FOUR

THE BUCHNERS TO THE WARBURG GROUP

Although a Bavarian, Eduard Buchner (1860–1917) was a loyal subject of the Hohenzollern Kaiser, and at the outbreak of World War I he felt obliged to join the German army. From 1888, when he received his doctoral degree in organic chemistry at Munich, to 1914, he held successive appointments at Kiel, Tübingen, Berlin, Breslau, and Würzburg, usually in analytical chemistry.[1] Before his elder brother Hans Buchner (1850–1902) was able to help him in 1884 enter the university Eduard had worked at a preserve and canning factory, where he may have acquired an interest in fermentation. As a student and teaching assistant in the famous chemical institute of Adolf von Baeyer (1835–1917), Buchner worked on diazo compounds, and continued to publish papers in that field until 1905. Early in his distinguished scientific career, Baeyer had published an interesting chemical explanation of the fermentative conversion of glucose into lactic acid.[2] This past interest may have been a factor in Baeyer's provision of funds in 1890 to set up a fermentation laboratory for Buchner. There, he tried, without success, to prepare an active cell-free yeast extract. In 1893 Buchner went to Kiel, and three years later he moved to Tübingen to be associate professor of analytical chemistry, but in 1898 he moved again, this time to Berlin, as professor of chemistry at the College of Agriculture. It was during the stay in Tübingen that Buchner participated with his brother Hans and Martin Hahn (1865–1934) in the work (at Hans Buchner's Munich Institute of Hygiene) which led to the successful preparation of a cell-free yeast extract which could ferment glucose. Eduard

[1] Harries, C. (1917). "Eduard Buchner," *Berichte der deutschen chemischen Gesellschaft* 50, pp. 1843–1876; Buchner, R. (1936). "Die politische und geistige Vorstellungswelt Eduard Buchners," *Zeitschrift für bayerische Landesgeschichte* 26, pp. 631–645; Schriefers, H. (1970), *Dictionary of Scientific Biography* 2, pp. 560–563. New York: Scribners.

[2] Baeyer, A. (1870). "Ueber die Wasserentziehung und ihre Bedeutung für das Pflanzenleben und die Gährung," *Berichte der deutschen chemischen Gesellschaft* 3, pp. 63–75.

was the sole author of the first (preliminary) publication in a scientific journal.[3] The paper had been received for publication on 11 January 1897, and before it appeared several months later, Eduard had spoken of the results on 4 February in his obligatory address on becoming an associate (*ausserordentlicher*) professor at Tübingen, and on 16 March, when Hans spoke to the Munich Morphological-Physiological Society. The oft-quoted conclusion in Eduard Buchner's published paper was:

> . . . the initiation of the fermentation process does not require so complicated an apparatus as is represented by the yeast cell. The agent responsible for the fermenting action of the press juice is rather to be regarded as a dissolved substance, doubtless a protein; this will be denoted *zymase*.[4]

The sole reference to the Munich institute is in the final sentence: "It has been found that the above hydraulic press method is suitable for the preparation of the contents of bacterial cells, and experiments to this end, including pathogenic bacteria, are in progress at the hygienic institute in Munich."[5]

It is well known that the discovery attributed to Eduard Buchner, and for which he received the 1907 Nobel Prize in Chemistry, was a happy accident in which Hans Buchner's efforts to prepare bacterial extracts of medical value, the experimental skill of his assistant Martin Hahn, and Eduard's presence in Munich during the autumn vacation of 1896 all played a part. It is therefore dismaying to find that neither Hans Buchner nor Hahn are mentioned in the 1897 paper, and the sole indication of the role of the Munich institute is in the final sentence cited above. In his lecture on 16 March 1897, Hans Buchner stated:

> Eduard Buchner had already grasped the idea several years ago, and realized in practice, the preparation of the cell juice of lower fungi, especially yeast cells, through mechanical maceration of the latter by the addition of sand. This procedure was recently improved by subjecting the macerated fungal cells to a pressure of 400–500 atmospheres . . . The press juice of the yeast cells soon showed a highly remarkable phenomenon, and it is to the credit of Eduard Buchner

---

[3] Buchner, E. (1897). "Alkoholische Gährung ohne Hefezellen (Vorlaufige Mitteilung)," *Berichte der deutschen chemischen Gesellschaft* 30, pp. 117–124.
[4] *Ibid.*, pp. 119–120.
[5] *Ibid.*, p. 124.

> that he was the first to interpret correctly, and to demonstrate experimentally, that the press juice was able to effect alcoholic fermentation, to cleave fermentable sugars to alcohol and carbonic acid.[6]

A further historical note was added by Martin Hahn, after the award of the Nobel Prize:

> When at the request of Hans Buchner, as his assistant, and the help of then candidate in pharmacy, now pharmacist Dr. Bullenheimer, who was E. Buchner's assistant and deputy for the maceration experiments, we carried out the maceration of the yeast with quartz sand according to the procedure used by E. and H. Buchner in 1893, we did not succeed in obtaining the yeast contents somewhat cell-free, in appreciable quantities or undiluted state. Only after I combined the addition of *kieselguhr* and the application of the hydraulic press with the maceration by quartz sand were there obtained large amounts of an almost cell-free fluid containing much protein. I then carried out animal tests, whereupon there appeared a rapid decomposition, that is the disappearance of coagulable protein, a phenomenon explicable by the action of an endotryptic enzyme. At the end of the summer semester, immediately before my departure on my vacation, I was able to lay before Prof. Hans Buchner the definitive results, with a precise quantitative statement (*Fixierung*) of the experimental conditions, under which a strongly protein-containing almost cell-free press juice could be obtained in large quantities. Of course, on this occasion my observation of the rapid decomposition of the press juice was mentioned, and we discussed the fact that the hitherto used antisepsis as the method of conservation for animal injection was rather ineffective and because of the resulting precipitates is unsuitable, and therefore upon resumption of the work one could examine conservation with salts, glycerine, or sugar. Prof. Eduard Buchner, who was regularly informed by his brother about the progress of the work, but because of his absence from Munich, until then could not participate actively in it, happened to come to the Institute a few days after my departure. Because of my absence for several months I do not know who of those present at the Institute and participants in the research added sugar to the press juice for conservation at the direction of Hans Buchner. In any case, Hans Buchner informed me sometime later of the discovery of cell-free fermentation by his brother Eduard Buchner.[7]

---

[6] Buchner, H. (1897), "Die Bedeutung der activen löslichen Zellprodukte für den Chemismus der Zelle," *Münchener medizinische Wochenschrift* 44, pp. 299–302 (299–300).

[7] Hahn, M. (1908). "Zur Geschichte der Zymaseforschung," *Münchener medizinische Wochenschrift* 55, pp. 515–516 (516). See Weindling, P. (1979). *From Bacteriology to Social Hygiene: The Papers of Martin Hahn*. Oxford: Wellcome Unit for the History of Medicine.

In his Nobel address, Eduard Buchner acknowledged his debt to Martin Hahn for the introduction of diatomaceous earth and the hydraulic press and for the discovery of the proteolytic enzyme in yeast.

According to Eduard Buchner, the method of maceration of microorganisms with sand was filed in 1893 for a patent, which was denied.[8] Earlier unsuccessful attempts to prepare an active cell extract in this manner were reported by Friedrich Wilhelm Lüdersdorff (1801–1886) and Carl Schmidt (1822–1894) during the 1840s, and by Denys Cochin (1851–1922) in 1880.[9]

In 1872, Maria Mikhailovna Manasseina (1841–1903) claimed to have shown that yeast subjected to strong heat (150–300°C) could still ferment sugar, and after the appearance of Buchner's announcement she asserted her priority in the demonstration that alcoholic fermentation was not dependent on living yeast.[10] Manasseina has recently emerged from obscurity on the centenary of her claim for priority.[11]

The reception of Eduard Buchner's announcement, after brief skepticism, was favorable. In his authoritative book on fermentation and soluble ferments, Joseph Reynolds Green (1848–1914) wrote:

> The preparation of such an extract as this presents considerable difficulty, the operation of grinding the yeast being very tedious. Since Buchner's results were published it has been successfully carried out by Delbrück in Berlin, and by the writer and by MacFadyen, Morris and Rowland in this country.[12]

---

[8] Buchner, E., H. Buchner, and M. Hahn (1903). *Die Zymasegährung*, p. 20. Munich: Oldenbourg. See Neubauer, A. (2000). "Die Entdeckung der zellfreien Gärung," *Chemie in unserer Zeit* 34, pp. 126–133.

[9] Lüdersdorff, F. W. (1846). "Ueber die Natur der Hefe," *Annalen der Physik und Chemie* 76, pp. 408–411; Schmidt, C. (1847). "Gährungsversuche," *Annalen der Chemie und Pharmacie* 61, pp. 168–174; Cochin, D. (1880). "Recherches du ferment alcoholique soluble," *Annales de Chimie* [5] 21, pp. 430–432.

[10] Manasseina, M. (1872), "Beiträge zur Kenntnis der Hefe und zur Lehre von der alkoholischen Gährung" in: *Mikroskopische Untersuchungen*, J. Wiesner (ed.), pp. 116–128. Stuttgart; "Zur Frage von der alkoholischen Gärung ohne lebende Hefezellen," *Berichte der deutschen chemischen Gesellschaft* 30, pp. 3061–3062.

[11] Kästner, I. (1996). "Kein Nobelpreis für Maria Manasseina. Ein Beitrag zur Geschichte der Biochemie" in: *Dilettanten und Wissenschaft*, E. Straus (ed.), pp. 123–134; Jaenicke, L. (2002). "Wer begründete die in-vitro-Enzymologie?" *Chemie in unserer Zeit* 36, pp. 64–65.

[12] Green, J. R. (1901). *The Soluble Ferments and Fermentation*. 2nd ed., p. 359. Cambridge University Press.

In a chemical review, Felix Ahrens (1863–1910) concluded: "The old argument: 'What is fermentation?' has ended: Fermentation is a chemical process."[13] The former Pasteur students were enthusiastic. Émile Roux wrote:

> It is the fact of the extraordinary adherence of some toxins to the microbes which produce them that put on the good path the extraction of the soluble alcoholic ferment. M. Eduard Buchner, of Munich, the brother of the well-known bacteriologist, had the idea of grinding the yeast with quartz sand to remove the cell envelopes, to add infusorial earth and then subjecting the resulting paste to a pressure of five hundred atmospheres in a hydraulic press.[14]

Jean Effront (1856–1931) wrote: "This discovery gives a definitive explanation for alcoholic fermentation; it will certainly have a great influence on the study of analogous phenomena, and will lead to the discovery of many other enzymes."[15]

There was also considerable discussion about the composition of zymase, and its difference from known soluble enzymes such as invertase. Buchner himself stated in his first 1897 paper that "zymase belongs to the true proteins and stands closer to the living protoplasm of the yeast cell than does invertin."[16] This statement suggests lingering adherence to Nägeli's theory of fermentation and to the concept of "living proteins" of Nägeli's disciple Oscar Loew (1844–1941).[17] However, when Richard Neumeister (1857–1906) suggested in 1897 that the fermentation was not effected by a single substance, but by a more complex set of cell constituents, Buchner replied that

> So long as no experimental data whatever can be offered in favor of this complicated hypothesis, it is provisionally expedient to adhere to the simpler assumption of a homogeneous zymase as the agent of fermentation. Neumeister's view appears to have arisen from the need to explain more easily the "complicated function" of zymase, whose action as a single substance seemed difficult to understand.[18]

---

[13] Ahrens, F. B. (1902). "Das Gährungsproblem," *Sammlung chemischer und chemisch-technischer Vorträge* 7, pp. 445–494 (494).

[14] Roux, E. (1898). "La fermentation alcoolique et l'évolution de la microbie," *Revue Scientifique* 27, pp. 833–840 (838).

[15] Effront, J. (1899). *Les Enzymes et leurs Applications*, p. 318. Paris: Cabre & Naud.

[16] Buchner, E. (1897) (note 274), p. 120.

[17] Loew, O. (1896). *The Energy of Living Protoplasm.* London: Kegan Paul et al.

[18] Buchner, E. (1903) (note 279), p. 38.

Later, the physiologist Max Rubner (1854–1932) emphasized the fact that the cell-free extract was much less effective in fermenting sugar than intact yeast, and maintained that although a small part of yeast fermentation might be caused by Buchner's zymase, the major role was played by the "living proteins" in the protoplasmic structure.[19] Many of Buchner's scientific papers, including his last one (in 1914) were defenses of his position. As Jacques Loeb (1859–1924) expressed the prevailing view, which led to the award of the Nobel Prize,

> Through the discovery of Buchner, Biology was relieved of another fragment of mysticism. The splitting up of sugar into CO_2 and alcohol is no more the effect of a "vital principle" than the splitting up of cane sugar by invertase. The history of this problem is instructive, as it warns us against considering problems as beyond our reach because they have not yet found their solution.[20]

An excellent account of the immediate reception of the discovery of cell-free fermentation has been presented by the historian Robert Kohler.[21] He has also offered the hypothesis that this discovery was an important event, along with the finding of "oxidases" and "synthetases," in the "origin of biochemistry."[22] The interest of historians in "origins" or "revolutions" is understandable, as is their interest in Nobel Prize winners, but in my view what came before Buchner's zymase and afterwards far outshines what he contributed to the solution of the problem of the nature of alcoholic fermentation.

As is not uncommon, a difficult experimental procedure is soon replaced by a better one. After several improvements by Buchner and others what eventually emerged in 1911 was the method of Aleksandr Nikolaevich Lebedev (1881–1958) for the extraction of zymase by means of water from the macerated dried yeast cells.[23] It was also learned that whereas the brewers bottom yeast available in Munich usually gave a good yield of zymase, the top yeast from a

---

[19] Rubner, M. (1913). "Die Ernährungsphysiologie der Hefezelle bei der alkoholischen Gärung," *Archiv für Physiologie* Supplement, pp. 1–392 (55).

[20] Loeb, J. (1906). *The Dynamics of Living Matter*, p. 22. New York: Columbia University Press.

[21] Kohler, R. E. (1971). "The reception of Eduard Buchner's discovery of cell-free fermentation," *Journal of the History of Biology* 5, pp. 327–353.

[22] Kohler, R. E. (1973). "The enzyme theory and the origin of biochemistry," *Isis* 64, pp. 181–196.

[23] Lebedev, A. (1912). "Extraction de la Zymase par simple maceration," *Annales de L'Institut Pasteur* 28, 8–37.

Paris brewery or one in England often gave inactive maceration extracts. Variants of the preparation of active dried zymase extracts (termed "zymin") continued to appear until at least 1942.[24]

One of Eduard Buchner's first competitors was Augustyn Wróblewski (1866–?), a Polish investigator in Cracow, who disappeared during the 1914–1918 war. He reported in 1901 that the addition of inorganic phosphate to a yeast juice caused a marked increase in the rate of fermentation as measured by the CO_2 release. He interpreted this phenomenon as due to a control of the acid-base balance (the German equivalent of "buffer" was not yet available).[25] Wróblewski's observation was confirmed by Arthur Harden (1865–1940) and his associate William John Young (1878–1947) as well as by Leonid Aleksandrovich Ivanov (1871–1962), who demonstrated the formation, as possible intermediates, of organic phosphorus compounds during alcoholic fermentation.[26]

In his 1907 Nobel address, Buchner did not mention these observations, and spoke of his view that "lactic acid plays an important role in the cleavage of sugar and probably appears as an intermediate in alcoholic fermentation."[27] He proposed that the term zymase be applied to the enzyme that cleaves glucose to lactic acid, and that the conversion of lactic acid to alcohol and CO_2 is effected by another enzyme, named "lactacidase." As a former student in Adolf Baeyer's institute, Buchner knew of Baeyer's 1870 paper, in which he suggested an "accumulation" of oxygen at the center of the linear six-carbon chain of glucose with the formation of two lactic acid molecules. Baeyer's concluding statement was:

> It is assumed in the above reflections that the molecules of sugar which undergo fermentation do not serve as nutrients of the yeast, but only receive an impulse comparable to the action of heat and dehydrating

[24] Neuberg, C. and H. Lustig (1942). "Preparation of active zymase extracts from top yeast," *Archives of Biochemistry* 2, pp. 191–196.

[25] Wróblewski, A. (1901). "Ueber den Buchner'schen Hefepresssaft," *Journal für praktische Chemie* [2] 64, pp. 1–70.

[26] Harden, A. and W. J. Young (1905). "The alcoholic ferment of yeast juice," *Proceedings of the Royal Society* 77B, pp. 405–420; Ivanov, L. (1906). "Ueber die Synthese der phospho-organischen Verbindungen in abgetöteten Hefezellen," *Zeitschrift für physiologische Chemie* 50, pp. 281–288.

[27] Buchner, E. and J. Meisenheimer (1904). "Die chemischen Vorgänge bei der alkoholischen Gärung," *Berichte der deutschen chemischen Gesellschaft* 37, pp. 417–428 (420–421).

Glycerol	Glyceraldehyde	Dihydroxyacetone	Pyruvic acid	Acetaldehyde
CH_2OH–CH(OH)–CH_2OH	CHO–CH(OH)–CH_2OH	CH_2OH–C=O–CH_2OH	COOH–C=O–CH_3	CHO–CH_3

Glycerol ←(± 2H)→ Glyceraldehyde → Dihydroxyacetone

Pyruvic acid —($-CO_2$)→ Acetaldehyde

Pyruvic acid ↕ ± 2H ↕ Lactic acid; Acetaldehyde ↕ ± 2H ↕ Ethanol

Glyceric acid	Methyl glyoxal	Lactic acid	Ethanol
COOH–CH(OH)–CH_2OH	CHO–C=O–CH_3	COOH–CH(OH)–CH_3	CH_2OH–CH_3

Glyceraldehyde —(± O)→ Glyceric acid; Glyceraldehyde —($\pm H_2O$)→ Methyl glyoxal → Lactic acid

Trioses related to glyceraldehyde and pyruvic acid

> agents. Liebig's latest researches on fermentation confirm this assumption and thereby give further evidence for proposed views about the chemical process in fermentatin.[28]

It should be noted that Baeyer did not assume that lactic acid is an intermediate in alcoholic fermentation; for the formation of alcohol and CO_2 he specified cleavages at three carbon-carbon bonds of the hexose.

During the nineteenth century lactic acid assumed importance in muscle physiology, and the "lactic fermentation" in animal tissues was termed "glycolysis." Richard Neumeister, who was skeptical about Buchner's zymase remarked:

> The formation of lactic acid evidently occurs as a consequence of the interaction of certain proteins present in the living muscle plasma, and is certainly no less complex a process than the cleavage of sugar to alcohol and carbon dioxide in the yeast press juice. What we understand by enzymes could not play a role in either case.[29]

It should be recalled that the connection between muscular contraction and the formation of lactic acid was definitely established

---

[28] Baeyer, A. (1870) (note 273), p. 75.

[29] Neumeister, R. (1897). "Bemerkungen zu Eduard Buchners Mitteilungen," *Berichte der deutschen chemischen Gesellschaft* 30, pp. 2963–2966 (2965).

only in 1907 by the work of Walter Morley Fletcher (1873–1933) and Frederick Gowland Hopkins (1861–1947).[30]

The lactic acid theory of alcoholic fermentation was short lived. In 1906 Arthur Slator (1879–1953) pointed out that lactic acid is fermented poorly, if at all, by brewers yeast, and that for a substance to be an intermediate in a process in which glucose is fermented, its rate of fermentation must be no lower than that of glucose.[31] Attention then shifted to the possible role of other three-carbon compounds such as glyceraldehyde and methyl glyoxal, of which more later.

After his undergraduate chemical studies with Henry Roscoe (1833–1915) and Carl Schorlemmer (1834–1892) at Owens College, Manchester, in 1888 Harden received his Ph.D. in organic chemistry at Erlangen. He then was a teacher in Manchester until 1897, when he went to what became in 1903 the Lister Institute, where he conducted research on alcoholic fermentation the rest of his life.[32] Harden began this research at the suggestion of Allen Macfadyen (1860–1907), the director of the institute and a bacteriologist who was impressed by Hans Buchner's program of macerating bacterial cells in order to extract antitoxins, and by Eduard Buchner's discovery of zymase.[33]

In 1903 Harden reported that the addition of blood serum to a mixture of Buchner's yeast juice increased the rate of CO_2 formation and he attributed this finding to "an inhibitory effect which the serum exerts on the proteolytic enzyme of the press-juice; one may therefore infer that the agent responsible for alcoholic fermentation is active for a longer time."[34]

---

[30] Fletcher, W. M. and Hopkins, F. G. (1907). "Lactic acid in amphibian muscle," *Journal of Physiology* 35, pp. 247–309; Hopkins, F. G. (1921). "The chemical dynamics of muscle," *Johns Hopkins Hospital Bulletin* 32, pp. 359–367.

[31] Slator, A. (1906), "Studies in fermentation. Part I: The chemical dynamics of alcoholic fermentation by yeast," *Journal of the Chemical Society* 89, pp. 128–142.

[32] Hopkins, F. G. and C. J. Martin (1942). "Arthur Harden (1865–1940)," *Obituary Notices of Fellows of the Royal Society* 4, pp. 3–14; Manchester, K. L. (2000). "Arthur Harden as unwitting pioneer of metabolic control analysis," *Trends in Biochemical Sciences* 25, pp. 89–92. See also Chick, H. et al. (1971). *War on Disease. A History of the Lister Institute*. London: Andre Deutsch.

[33] Kohler, R. E. (1974). "The background of Arthur Harden's discovery of cozymase," *Bulletin of the History of Medicine* 48, pp. 22–40.

[34] Harden, A.(1903). "Ueber alkoholische Gährung mit Hefe-Pressstoff (Buchners Zymase) bei Gegenwart von Blutserum," *Berichte der deutschen chemischen Gesellschaft* 36, pp. 715–716 (716).

In continuing the research, he soon found that the effect he observed was not related to the inhibition of proteolysis, but that the stimulation of fermentation was attributable to the presence in the serum of inorganic phosphate. Moreover, he found that "the fermentation of glucose is dependent upon the presence of a dialyzable substance which was not destroyed by heat."[35] This substance was named a "co-ferment," (a term first used in 1897 by Gabriel Bertrand (1867–1962)), but was renamed "cozymase" in 1923 by Hans Euler-Chelpin (1873–1964), who devoted many years to its isolation and the elucidation of its chemical structure. The task was only completed during the 1930s. Euler (the name in scientific writings) and Harden shared the 1929 Nobel Prize in Chemistry. It is noteworthy that in the first three editions (1911, 1918, 1923) of his book *Alcoholic Fermentation*, Harden expressed uncertainty about the role of the hexose diphosphate he and Young had discovered. For example, in the third edition, he stated that "It is not impossible that the hexose phosphate is formed by combined synthesis and esterification from smaller groups produced by the rupture of the sugar molecule."[36] As his biographers put it: "He mistrusted the use of his imagination beyond a few paces in advance of the facts. Had he exercised less restraint, he might have gone further; as it was he had little to withdraw."[37] Harden and Young did venture to suggest two equations to account for their observations:

$$(1)\ 2\ C_6H_{12}O_6 + 2\ Na_2HPO_4 = C_6H_{10}O_4(PO_4Na_2)_2 + 2\ H_2O + CO_2 + 2\ C_2H_6O$$

$$(2)\ C_6H_{10}O_4(PO_4Na_2)_2 + 2\ H_2O = C_6H_{12}O_6 + 2\ Na_2HPO_4$$

In the presence of arsenate, reaction (2) supplies phosphate at a rate that is sufficient to maintain reaction (1) at maximal velocity. In 1914, however, Harden and Robert Robison (1883–1941) found in fermenting yeast juice a hexose monophosphate; after World War I Robison described it in 1922, as a mixture of glucose and fructose monophosphates.[38]

---

[35] Harden, A. and W. J. Young (1906), p. 410.

[36] Harden, A. (1923). *Alcoholic Fermentation.* 3rd ed., p. 109. London: Longmans, Green.

[37] Hopkins, F. G. and C. J. Martin (1942), p. 14.

[38] Robison, R. (1922). "A new phosphoric ester produced by the action of yeast juice on hexoses," *Biochemical Journal* 16, pp. 809–824.

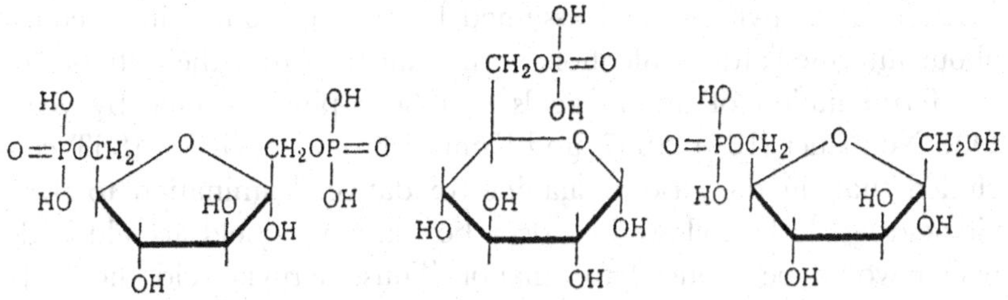

Fructose-1,6-diphosphate Glucose-6-phosphate Fructose-6-phosphate

After the demise of lactic acid as an intermediate in yeast fermentation, Alfred Wohl (1863–1939), professor of chemistry at the Danzig *Technische Hochschule*, proposed a modification of Baeyer's hypothesis. In Wohl's scheme glucose (in its enol form) is cleaved to methyl glyoxal and glyceraldehyde:[39]

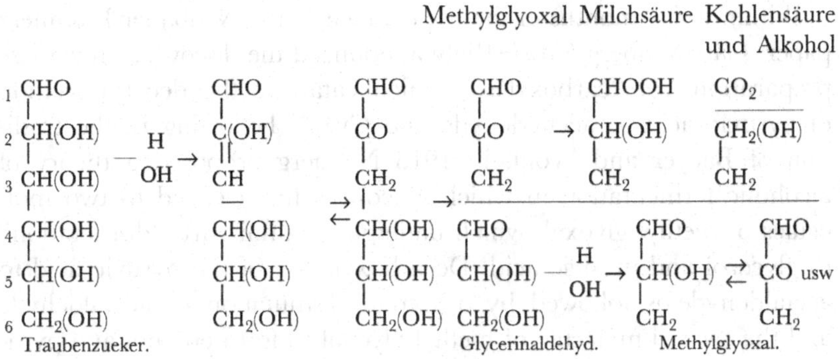

```
                                             Methylglyoxal  Milchsäure  Kohlensäure
                                                                        und Alkohol
1 CHO                 CHO       CHO          CHO            CHOOH       CO2
2 CH(OH)     H     →  C(OH)     CO           CO          →  CH(OH)      CH2(OH)
3 CH(OH)     OH       CH        CH2          CH2            CH2         CH2
                               →
4 CH(OH)              CH(OH)  ← CH(OH)   CHO                CHO              CHO
5 CH(OH)              CH(OH)    CH(OH)   CH(OH)   H   →     CH(OH)    ⇄      CO  usw
                                                  OH
6 CH2(OH)             CH2(OH)   CH2(OH)  CH2(OH)            CH2              CH2
  Traubenzueker.                         Glycerinaldehyd.   Methylglyoxal.
```

Wohl's scheme of alcoholic fermentation

Several investigators, including Buchner, tested these three-carbon compounds, along with dihydroxyacetone for their fermentability by yeast juice. Because he found dihydroxyacetone to be fermented more rapidly than the other two compounds, Buchner inserted it into his theory in place of lactic acid, and assumed the existence of some enzyme other than "lactacidase."

[39] Wohl, A. (1907). "Die neueren Ansichten über den chemischen Verlauf der Gärung," *Biochemische Zeitschrift* 5, pp. 45–64.

After 1910, pyruvic acid assumed larger importance in discourse about intermediates in alcoholic fermentation. From their studies on the fermentation of amino acids such as a-phenylglycine by yeast, Otto Neubauer (1874–1957) and Konrad Fromherz (1883–1963) concluded that, in the case of alanine, oxidative deamination to pyruvic acid would be followed by decarboxylation to yield acetaldehyde, which would be reduced to ethanol. Thus, pyruvic acid should be readily fermentable. Neubauer added the statement:

> Our own experiments, which are not fully completed, have confirmed this conclusion. The thought follows that pyruvic acid could be an intermediate in the alcoholic fermentation of sugar . . . I ask colleagues to leave to us the further study of the role of pyruvic acid in the fermentation of sugar; also, it is intended to study the question whether it is an intermediate in the combustion of sugar in higher animal organisms.[40]

Needless to add, this permission was not granted.[41]

Within a few months of the appearance of the Neubauer-Fromherz paper, Carl Neuberg (1877–1956) announced the discovery, in zymase preparations, of "carboxylase" which catalyzes the decarboxylation of pyruvic acid to acetaldehyde and CO_2.[42] Following in the tradition of Baeyer and Wohl, in 1913 Neuberg proposed a theory of alcoholic fermentation in which glucose is first cleaved to two molecules of methyl glyoxal, which undergo a Cannizzaro "dismutation" to glycerol and pyruvic acid. Decarboxylation of the pyruvic acid to acetaldehyde is followed by a second dismutation of acetaldehyde and the second molecule of methyl glyoxal to form ethanol and pyruvic acid:[43]

---

[40] Neubauer, O. and K, Fromherz (1911). "Über den Abbau der Aminosauren bei der Hefegärung," *Zeitschrift für physiologische Chemie* 70, pp. 326–350 (350).

[41] Fernbach, A. and M. Schoen (1913). "L'acide pyruvique, produit de la vie de la levure," *Comptes Rendus* 157, pp. 1478–1480.

[42] Neuberg, C. and L. Karczag (1911). "Über zuckerfreie Hefegärungen, IV, Carboxylase, ein neues Enzym der Hefe," *Biochemische Zeitschrift* 36, pp. 68–75.

[43] Neuberg, C. and J. Kerb (1914). "Über zuckerfreie Hefegärungen. XIII. Zur Frage der Aldehydbildung bei der Gärung von Hexosen sowie bei der sog. Selbstgärung," *Biochemische Zeitschrift* 58, pp. 158–170.

α) $C_6H_{12}O_6 - 2\ H_2O = C_6H_8O_4$ (Methylglyoxal-aldol).

β) $C_6H_8O_4 = 2\ CH_2 : C(OH).\ CHO$ bzw . $2\ CH_3 .\ CO .\ COH$ (Methylglyoxal).

γ) $CH_2 : C(OH) .\ COH + H_2O + \overset{H_2}{\underset{O}{|}} = \begin{matrix} CH_2OH .\ CHO .\ CH_2OH \text{ (Glycerin)} \\ + \\ CH_2 : C(OH) .\ COOH \text{ (Brenztraubensäure)} \end{matrix}$

δ) $CH_3 .\ CO .\ COOH = CO_2 + CH_3 .\ COH$ (Acetaldehyd)

ε) $\begin{matrix} CH_3 .\ CO .\ COH \\ CH_3 .\ COH \end{matrix} + \overset{O}{\underset{H_2}{|}} = \begin{matrix} CH_3 .\ CO .\ COOH \text{ (Brenztraubensäure)} \\ CH_3 .\ CH_2OH \text{ (Äthylalkohol).} \end{matrix}$

Neuberg's scheme of alcoholic fermentation (1913)

In this theory there was no place for the phosphorylated compounds of the kind identified by Harden and Young and by Ivanov; the fact that methyl glyoxal is not fermented by yeast was explained by assuming that one of its isomers is the "true" intermediate. An argument in favor of methyl glyoxal was the discovery by Henry Drysdale Dakin (1880–1952) and Harold Ward Dudley (1887–1935) of a widely distributed enzyme (named "glyoxalase") which catalyzes the interconversion of methyl glyoxal and lactic acid.[44]

In adopting the Cannizzaro reaction, Neuberg followed Jacob Karol Parnas (1884–1949), who reported in 1910 that an "aldehyde mutase" present in animal tissues can catalyze the conversion of an aldehyde into a mixture of the corresponding alcohol and acid:

$$2\ RCHO + H_2O = RCH_2OH + RCOOH$$

Stanislao Cannizzaro (1826–1910) had shown in 1853 that this reaction is promoted by alkali. Parnas wrote:

> In the Cannizzaro rearrangement of the aldehydes we have come to know a simple system of coupled reactions, in which through oxygen transfer and hydrogen uptake there occur simultaneous oxidation and reduction. Through an enzyme of the liver the reaction is catalyzed to such an extent that it leads to the complete disappearance of the

[44] Dakin, H. D. and Dudley, H. W. (1913). "Glyoxylase. III. The distribution of the enzyme and its relation to the pancreas," *Journal of Biological Chemistry* 15, pp. 463–474.

> aldehydes . . . Aldehydes may be regarded as general reductants for the reduction of carbonyl groups in the animal organism. Through specific ferments the Cannizzaro reaction is accelerated, and there are formed an alcohol (or a hydroxy acid) and a fatty acid.[45]

A similar enzyme (named "aldehydase") was found by Federico Battelli (1867–1941) and Lina Salomonovna Stern (1878–1968).[46]

In Berlin, Carl Neuberg attained a high position in pre-Nazi German science. After becoming *Privatdozent* in 1903 and *ausserordentlicher* professor in 1906, in 1913 he was appointed head of the biochemistry section in the newly established Kaiser Wilhelm Institute of Experimental Therapy, with August von Wassermann (1866–1925) as director. After Wassermann died, Neuberg became director of the entire institute, only to be forced in 1934 by the Nazis to resign the post. He left Germany in 1938, and spent his last years in New York City. Neuberg had edited the *Biochemische Zeitschrift*, published by Julius Springer in Berlin, since 1906; most of his large literary output, and that of other leading German biochemists, appeared in that journal.

During World War I, Neuberg's theory of alcoholic fermentation received support from its successful application to the manufacture of glycerol in Germany. He had shown that the addition of sodium sulfite to a yeast fermentation mixture "traps" acetaldehyde, thus decreasing the yield of ethanol and CO_2 in favor of glycerol and acetaldehyde:

> Glycerol is the reduction equivalent of pyruvic acid, which decomposes to carbonic acid and acetaldehyde. If the reduction of the latter is blocked, the only remaining possibility is the increased correlative formation of glycerol.[47]

The sulfite process was perfected during World War I by Wilhelm Connstein and Friedrich Lüdecke; in their published paper they stated that their experiments had begun in 1914, but

---

[45] Parnas, J. (1910). "Über fermentative Beschleunigung der Cannizzaroschen Aldehyd-umlagerung durch Gewebssäfte," *Biochemische Zeitschrift* 28, pp. 274–294. See Zelinska, Z. (1987). "Jakub Karol Parnas," *Acta Physiologica Polonica* 38, pp. 91–99.

[46] Battelli, F. and L, Stern (1910). "Die Aldehydrase in den Tiergeweben," *Biochemische Zeitschrift* 29, pp. 130–151.

[47] Neuberg, C. and Reinfurth, E. (1919). "Weitere Untersuchungen über die korrelative Bildung von Acetaldehyd und Glycerin bei der Zuckerspaltung und neue Beiträge zur Theorie der alkoholischen Gärung," *Berichte der deutschen chemischen Gesellschaft* 52, pp. 1677–1703 (1681).

> . . . could not be published earlier because, during the war, the German army administration had an interest in keeping the experiments and results secret. Our work arose from the necessity of the time and owes its origin to the expectation that the supply of glycerol available to the European Central Powers would be insufficient, because of the blockade.[48]

Neuberg later offered other schemes of fermentation. One led to the production of equivalent amounts of pyruvic acid and glycerol, the other to methyl glyoxal.[49] He also found in yeast fermentation mixtures a hexose monophosphate, which was duly named Neuberg-ester, to distinguish it from the one found by Robison. Neuberg was nominated several times for the Nobel Prize in Physiology or Medicine, but during 1920–1933 methyl glyoxal lost most of its attraction, largely because of the outstanding work of Otto Meyerhof (1884–1951) and Gustav Embden (1884–1933).[50]

In adopting the concept of the Cannizzaro dismutation of aldehydes as an oxidation-reduction process, Neuberg brought the discussion of the mechanism of alcoholic fermentation into the arena of controversy during 1910–1930 about the existence or role of oxidative and reductive enzymes. The leading figure in the dispute was Otto Heinrich Warburg (1883–1970); he was opposed by Heinrich Otto Wieland (1877–1957) and Torsten Ludvig Thunberg (1873–1962).

Son of the noted professor of physics Emil Warburg (1846–1931), a recent recipient of a Ph.D. in organic chemistry for work with Emil Fischer, and of an M.D. in Heidelberg, during 1908–1914 Otto Warburg began research on the effect of cyanide and ethyl urethane on the oxygen uptake by sea urchin eggs and red blood cells. For this purpose, he greatly modified the available manometric apparatus. In 1913, Warburg was appointed a member of the Kaiser Wilhelm Society, with an independent laboratory for his research, but soon after the outbreak of World War I, he joined a cavalry regiment of the German army, and remained in military service until October 1918. Warburg spent the rest of his life at his Kaiser Wilhelm

---

[48] Connstein, W. and Lüdecke, F. (1919). "Über Glyceringewinnung durch Gärung," *Berichte der deutschen chemischen Gesellschaft* 52, pp. 1385–1391 (1385).

[49] Neuberg, C. and Kobel, M. (1929). "Weiteres über die Vorgänge bei desmolytischen Bildung von Methylglyoxal durch Hefe," *Biochemische Zeitschrift* 210, pp. 466–488; (1930). "Die Zerlegung von nicht phosphoryliertem Zucker durch Hefe unter Bildung von Glycerin und Brenztraubensäure," *ibid.* 229, pp. 446–454.

[50] Björk, R. (2001). "Inside the Nobel Committee on Medicine: Prize competition procedures 1901–1950 and the case of Carl Neuberg," *Minerva* 39, pp. 393–408.

(later Max Planck) Institute.[51] Despite his partly Jewish ancestry, the Nazi authorities allowed Warburg to continue there after 1933, probably because of his social connections and his war record.

In an article published in 1914, Warburg concluded "that the oxygen respiration in the egg is an iron catalysis; that the oxygen consumed in the respiratory process is taken up initially by dissolved or adsorbed ferrous iron."[52] He developed this concept during the decade after the war, and numerous model experiments were performed in which oxidations were catalyzed by iron-containing charcoals; these were first prepared by the incineration of blood, and later of hemin or of impure aniline dyes contaminated with iron salts. Various amino acids (cystine, tyrosine, leucine) were extensively oxidized by oxygen in the presence of such charcoals, and the catalysis was inhibited by cyanide and ethyl urethane. The theory of cellular respiration Warburg offered in 1924 proposed a cyclic process in which

> ... molecular oxygen reacts with divalent iron, whereby there results a higher oxidation state of iron. The higher oxidation state reacts with the organic substrate with the regeneration of divalent iron ... Molecular oxygen never reacts directly with the organic substrate.[53]

To justify the use of charcoal models for a theory of physiological oxidation, Warburg stated:

> The experiments ... are model experiments in so far as the conditions under which we work are simpler than those in the cell. The experiments are more than model experiments if one succeeds with the help of iron in transferring the oxygen to the combustible substances of the cell.[54]

He believed that the results justified the reiteration of the view that he had expressed in 1914:

> Thus there arises the remarkable interplay of unspecific surface forces and specific chemical forces, characteristic of the hemin-charcoal as

[51] Krebs, H. A. (1972). "Otto Heinrich Warburg (1883–1970)," *Biographical Memoirs of Fellows of the Royal Society* 18, pp. 629–699.

[52] Warburg, O. (1914). "Über die Rolle des Eisens in der Atmung des Seeigeleies nebst Bemerkungen über einige durch Eisen beschleunigte Oxydationen," *Zeitschrift für physiologische Chemie* 92, pp. 231–256 (253–254).

[53] Warburg, O. (1924). "Über Eisen, den sauerstoffübertragenden Bestandteil des Atmungsferments," *Biochemische Zeitschrift* 152, pp. 479–494 (479).

[54] *Ibid.*, p. 483.

> well as the living substance. Both behave on the one hand like unspecific surface catalysts, on the other as specific metal catalysts. The specific anticatalyst is hydrocyanic acid, the unspecific anticatalysts are the narcotics.[55]

This article, entitled "On iron, the oxygen-transferring constituent of the respiratory enzyme (*Atmungsferment*)" appeared after a review article by Heinrich Wieland, in which he questioned the validity of Warburg's conclusions from experiments on the oxidation of amino acids in the presence of iron-charcoals:

> He sees the active agent in the iron content of the catalyst and believes in an activation of molecular oxygen by the metal. Since the reaction is inhibited by hydrocyanic acid, just as the action of the respiratory enzymes, Warburg believes it to be also necessary to attribute to iron the exclusive role in their action. He considers the inhibition by hydrocyanic acid to be due to the formation of ferrocyanide compounds. Because of the entirely different order of magnitude in enzyme action, the attempt to derive the function of iron-containing enzymes from the catalytic ability of inorganic lower (2)-oxides is inadmissible, as shown by Willstätter in his first paper on peroxidase.[56]

Wieland had studied chemistry at various German universities, and in 1901 received his Ph.D. in organic chemistry for work with Johannes Thiele (1865–1918) in Baeyer's Munich institute. Wieland remained in Munich until 1922 (*Privatdozent*, 1904; *ausserordentlicher* Professor, 1914). After three years as a professor at Freiburg, Wieland returned to Munich to succeed Richard Willstätter (1872–1942) as head of the chemical institute. He retired in 1950.[57] An investigator with wide research interests, in 1913 Wieland sought to apply to biological oxidations the results of his studies on the catalysis, by finely divided palladium, of the oxidation of compounds such as an aldehyde (RCHO) to an acid (RCOOH). He reported that this process did not involve molecular oxygen, but was a "dehydrogeneration" of the water adduct [$RCH(OH)_2$], and that dyes such as methylene blue could act as oxidants

---

[55] *Ibid.*, p. 488.

[56] Wieland, H. (1922). "Über den Mechanismus der Oxydationsvorgänge," *Ergebnisse der Physiologie* 20, pp. 477–518 (502).

[57] Witkop, B. (1992). "Remembering Heinrich Wieland: Portrait of an organic chemist and founder of modern biochemistry," *Medical Research Reviews* 12, pp. 195–274.

> ... the so-called reduction enzymes, often treated in the literature, lose their separate status if one can bring the proof that their obvious reduction action, for example the decolorization of a dye by means of a substrate, may also be used for the hydrogenation of the oxygen molecule, if one ... can show that the "reductase" can also function at the same time as an oxidase.[58]

Although Wieland's initial experimental results were later shown to depend on impurities in the finely divided palladium,[59] his theory stimulated Thunberg to develop, during 1917–1920, a valuable technique, for which he devised a special test tube. Thoroughly washed minced tissue (e.g. frog muscle) was suspended in a solution containing methylene blue, which was not decolorized by the washed tissue. After the tube had been evacuated to remove oxygen, solutions of organic compounds were tipped in, and the time required for the decolorization was noted. From the results, Thunberg concluded that there were separate dehydrogenases for lactic acid, succinic acid, malic acid, citric acid, a-ketoglutaric acid, glutamic acid, and alanine.[60]

In his riposte to Wieland's criticism, Warburg questioned the biological relevance of the experiments with palladium black, and his emphatic conclusion (in italics) about Thunberg's results was "*Methylene blue, quinone and similar substances do not act in the cell like molecular oxygen, but like molecular oxygen + iron, that is like activated oxygen.*"[61] In 1924, two investigators independently made a significant contribution to the debate. From experiments on the inhibition by cyanide of the oxidation of succinate to fumarate by washed muscle tissue, and the reversal of the inhibition by methylene blue, Albert Fleisch (1892–1973) concluded that "The activation of oxygen as well as the activation of hydrogen is necessary for the oxidation of succinic acid,"[62] and Albert Szent-Györgyi (1893–1986) wrote:

---

[58] Wieland, H. (1913). "Über den Mechanismus der Oxydationsvorgänge," *Berichte der deutschen chemischen Gesellschaft* 46, pp. 3327–3342 (3339).

[59] Gillespie, L. J. and T. H. Liu (1931). "The reputed dehydrogenation of hydroquinone by palladium black," *Journal of the American Chemical Society* 53, pp. 3969–3972.

[60] Thunberg, T. (1920). "Zur Kenntnis des intermediären Stoffwechsel und der dabei wirksamen Enzyme," *Skandinavisches Archiv der Physiologie* 40, pp. 1–91.

[61] Warburg, O. (1923). "Über die Grundlagen der Wielandschen Atmungstheorie," *Biochemische Zeitschrift* 142, pp. 518–523 (522). See P. Werner (1997). "Learning from an adversary? Warburg against Wieland," *Historical Studies in the Physical and Biological Sciences* 28, pp. 173–196.

[62] Fleisch, A. (1924). "Some oxidation processes of normal and cancer tissue," *Biochemical Journal* 18, pp. 294–311 (311).

> The KCN-inhibited oxidation of succinic acid by muscle tissue can be reversed by the addition of methylene blue. The explanation of this result is that in the oxidation of succinic acid a double mechanism is operative, a mechanism of hydrogen activation according to *Wieland* and a mechanism of oxygen activation according to Warburg. The biological oxidation of the acid comes about through the harmonious cooperation of both processes. Molecular oxygen is unable to oxidize activated hydrogen. Molecular hydrogen is not a hydrogen acceptor. If the oxygen activation according to Warburg poisoned by cyanide, the oxidation ceases since the activated hydrogen cannot be burned. A new connection between active hydrogen and molecular oxygen is created by methylene blue.[63]

The participants in this debate do not appear to have known of the contemporary work of William Mansfield Clark (1884–1964) on the oxidation-reduction of dye systems such as methylene blue-leucomethylene blue. He defined the reduction of the dyes as "the transfer of an electron pair accompanied or not accompanied by hydrogen ions according to the state of acid-base equilibrium in the solution."[64]

During the 1920s the status of enzymes such as "succinate oxidase" was unclear, and Warburg expressed his view as follows:

> If one calls oxidases ferments that transport molecular oxygen, then the extracts contain oxidases, and if one classifies the oxidases, as is customary in ferment chemistry, according to the observed actions, then one has in the extracts different oxidases, glucose oxidase, alcohol oxidase, indophenol oxidase, and so on. Strictly speaking, there are as many different oxidases as extraction experiments. If the extract-oxidases had been preformed in the cell, a single type of cell would contain innumerable oxidases. But the multiplicity of oxidases in the living cell would be in opposition to a sovereign principle in the living substance . . . Therefore, if many different oxidases have been found in extracts of a cell type, these were not ferments that were already present in the living cell, but rather products of the transformation and decomposition of a single homogeneous substance present in life.[65]

---

[63] Szent-Györgyi, A. (1924). "Über den Mechanismus der Succin- und Paraphenylendiamin-oxydation. Ein Beitrag zur Theorie der Zellatmung," *Biochemische Zeitschrift* 150, pp. 195–210 (209–210).

[64] Clark, W. M. (1925). "Recent studies on reversible oxidation-reduction in organic systems," *Chemical Reviews* 2, pp. 127–178 (171).

[65] Warburg, O. (1929). "Atmungsferment und Oxydasen," *Biochemische Zeitschrift* 214, pp. 1–3.

I interrupt here the account of Warburg's contributions to note that during the 1920s he began his studies on the aerobic glycolysis of tumor tissues and on the quantum yield in photosynthesis, areas of research that also involved him in controversy. What stands out most strikingly, however, was the work with his remarkable assistant Erwin Negelein (1897–1979) to determine the photochemical "action spectrum" of the *Atmungsferment*. It involved measurements of the relative efficiency of various wavelengths of light in counteracting the inhibition, by carbon monoxide, of the respiration of yeast.[66]

In 1918, Otto Meyerhof reported the occurrence in animal tissues of a material seemingly identical with the co-ferment found in yeast by Harden and Young.[67] This discovery marks Meyerhof's entry into the study of the pathway and energetics of the conversion of glycogen to lactic acid in mammalian muscle. He had received his M.D. in 1909 at Heidelberg, and during 1910–1912 worked in Ludwig Krehl's clinic, where he gained the friendship of Otto Warburg. During the succeeding ten years, Meyerhof was in Kiel, where his work on the energetics of muscle glycolysis in 1922 won him a Nobel Prize in Physiology or Medicine, which he shared with Archibald Vivian Hill (1886–1977). This happy development brought him an appointment at the Kaiser Wilhelm Institute of Biology (1924–1929), and he was made head of a new Kaiser-Wilhelm Institute in Heidelberg (1929–1938). As a Jew, he was obliged to flee, and he came to Philadelphia via Paris, Spain and Portugal.[68]

Meyerhof's chief competitor in the search for the pathway of muscle glycolysis was Gustav Embden, whose untimely death in 1933 cut short a distinguished research career. After receiving his M.D. at Strassburg in 1899, Embden was associated with Franz Hofmeister (1850–1922), Hoppe-Seyler's successor as professor of physiological chemistry at that university. From Hofmeister, and Hofmeister's assistant Albrecht Bethe (1874–1954), Embden derived a stimulus to study what came to be called the "intermediate metabolism" of fatty acids,

[66] Warburg, O. and E. Negelein (1929). "Über das Absorptionsspectrum des Atmungsferments," *Biochemische Zeitschrift* 214, pp. 64–100.

[67] Meyerhof, O. (1918). "Über das Gärungscoferment im Tierkörper," *Zeitschrift für physiologische Chemie* 102, pp. 1–32.

[68] Muralt, A. von (1952). "Otto Meyerhof," *Ergebnisse der Physiologie* 47, pp. i–xx; Weber, H. H. (1972). "Otto Meyerhof – Werk und Persönlichkeit" in: *Molecular Energetics and Macromolecular Biochemistry*, H. H. Weber (ed.), pp. 3–13. Berlin: Springer.

amino acids, and carbohydrates. In 1904 Embden was appointed head of a newly established chemical laboratory at the Noorden clinic in Frankfurt am Main. Ten years later, Embden's laboratory became part of the new University of Frankfurt, and he was named Professor of Vegetative Physiology, a post he occupied until his death.[69]

In his initial studies on muscle glycolysis, Embden identified a precursor of lactic acid as "lactacidogen"[70] and he also suggested the following pathway:[71]

d-Glucose ⇆ active glyceraldehyde ⇆ d-lactic acid ⇆ pyruvic acid ⇆ acetaldehyde ⇆ alcohol

After World War I, he isolated the osazone of hexose diphosphate from muscle press juice in the presence of fluoride, and the barium salt of a hexose monophosphate from normal muscle.[72] Parallel work by Meyerhof's associate Karl Lohmann (1898–1978) showed that the hexose monophosphate in muscle is a mixture of the Robison and Neuberg esters (glucose-6-phosphate and fructose-6-phosphate).[73] It should be emphasized that as late as 1930, not only Harden, but also Meyerhof, were uncertain about the status of the isolated hexose phosphates as intermediates because they were fermented by yeast juice more slowly than glucose.

The studies of the Embden group during the mid-1920s on the phosphate compounds in muscle led to the isolation of an adenylic acid different from the adenosine-3-phosphate obtained on alkaline hydrolysis of yeast ribonucleic acid. The difference was evident from the resistance of the adenylic acid from yeast nucleic acid to the action of an enzyme which readily deaminated muscle adenylic acid

---

[69] Deuticke, H. J. (1935). "Gustav Embden," *Ergebnisse der Physiologie* 33, pp. 32–49; Cori, C. F. (1983). "Embden and the glycolytic pathway," *Trends in Biochemical Sciences* 8, pp. 257–259.

[70] Embden, G., F. Kalberlah, and H. Engel (1912). "Über Milchsäurebildung im Muskelpresssaft," *Biochemische Zeitschrift* 45, pp. 45–62.

[71] Embden, G. and M. Oppenheimer (1912). "Über den Abbau der Brenztraubensäure im Tierkörper," *Biochemische Zeitschrift* 45, pp. 186–206 (202).

[72] Embden, G. and M. Zimmermann (1927). "Über die Chemie des Lactacidogens. 5. Mitteilung," *Zeitschrift für physiologische Chemie* 167, pp. 114–136.

[73] Lohmann, K. (1928). "Über die Isolierung verschiedener natürlicher Phosphorsäureverbindungen und die Frage ihrer Einheitlichkeit," *Biochemische Zeitschrift* 194, pp. 306–307.

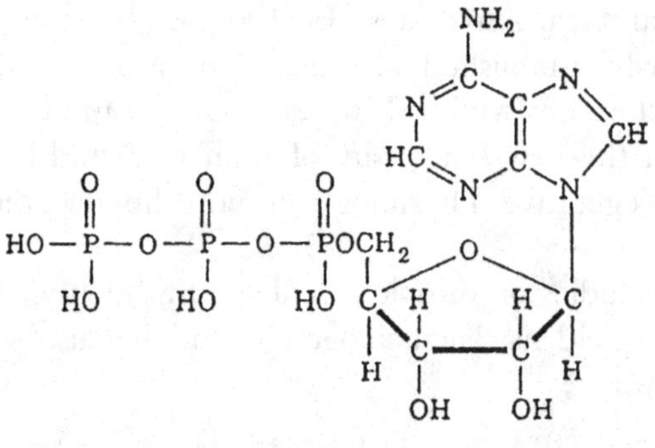

$$\text{Adenosine triphosphate structure}$$

Adenosine triphosphate (ATP)

to inosinic acid, long known to be hypoxanthine-5-phosphate.[74] At the time, this deamination seemed to be closely related to muscular contraction, but attention soon shifted to the finding by Cyrus Hartwell Fiske (1890–1978) and Yellapregrada Subbarow (1895–1948) of acid labile inorganic pyrophosphate in muscle, and Lohmann's demonstration that the pyrophosphate was attached to adenosine-5-phosphate to form adenosine triphosphate (ATP).[75]

In 1924, Embden had reported a delayed output of lactic acid during muscular contraction, and had suggested that the immediate energy for anaerobic muscular work was derived from some source other than lactic acid.[76] This observation was inconsistent with Meyerhof's widely accepted view that

> The first phase, the formation of lactic acid from carbohydrate, is anaerobic and spontaneous. This process is the immediate source of muscular force. In the second phase, with the expenditure of oxidation energy, the lactic acid is reconverted to carbohydrate. This second process corresponds to the recovery or restitution of the muscle.[77]

---

[74] Embden, G. and G. Schmidt (1929). "Über Muskeladenylsäure und Hefeadenylsäure," *Zeitschrift für physiologische Chemie* 181, pp. 130–139.

[75] Fiske, C. H. and Y. Subbarow (1927). "The nature of the 'inorganic phosphate' in involuntary muscle," *Science* 65, pp. 401–403; Lohmann, K. (1935). "Konstitution der Adenylpyrophosphorsäure und Adenosindiphosphorsäure," *Biochemische Zeitschrift* 282, pp. 120–123.

[76] Embden, G. (1924). "Untersuchungen über den Verlauf der Phosphorsäuren und Milchsäure bei der Muskeltätigkeit," *Klinische Wochenschrift* 3, pp. 1393–1396.

[77] Meyerhof, O. (1925). "Über den Zusammenhang der Spaltungsvorgänge mit der Atmung in der Zelle," *Berichte der deutschen chemischen Gesellschaft* 58, pp. 991–1001 (995).

According to A. V. Hill, Embden's "claim was not accepted, although it proved ultimately to be right."[78] Moreover, in 1927 Philip Eggleton (1903–1954) and Marion Grace Eggleton (1901–1970) stated:

> There is present in the skeletal muscle of the frog an organic phosphate compound which has hitherto been confused with inorganic phosphate owing to its rapid hydrolysis in acid solution to phosphoric acid. There may be more than one such compound, but the hypothesis of a single compound is sufficient to explain the available facts. We have given the name "phosphagen" to this substance. The results quoted in this paper established the fact that muscular contraction is accompanied by the removal of phosphagen, and subsequent recovery in oxygen is characterized by a rapid restitution of the phosphagen – a phase of recovery apparently independent of the relatively slow oxidative removal of lactic acid.[79]

Shortly afterward, Fiske and Subbarow showed "phosphagen" to be creatine phosphate.[80]

What A. V. Hill termed "the revolution in muscle physiology" was completed in 1930 by Einar Lundsgaard (1899–1968), who showed that the administration of iodoacetic acid to an animal abolishes the production of lactic acid by the muscles without abolishing their contractility. He concluded that phosphagen is the direct energy generating substance in muscular contraction, while the production of lactic acid effects the continual resynthesis of the cleaved phosphagen.[81]

A year before his sudden death in 1933, Embden (with his associates Deuticke and Kraft) published a preliminary scheme of muscle glycolysis that profoundly influenced the further development of the field. In introducing the scheme he stated:

> Recently, in the course of experiments designed for an entirely different purpose, i.e., the effects of ions on hexose phosphate synthesis, we were fortunate to observe an abundant formation of a beautifully crystallizing barium salt which could be identified as the secondary barium salt of a monophosphate ester of *l*-glyceric acid.[82]

---

[78] Hill, A. V. (1932). "The revolution in muscle physiology," *Physiological Reviews* 12, pp. 56–67 (57).

[79] Eggleton, P. and M. G. Eggleton (1927). "The physiological significante of 'phosphagen'," *Journal of Physiology* 63, pp. 155–161 (159).

[80] Fiske, C. H. and Y. Subbarow (1949). "Phosphocreatine," *Journal of Biological Chemistry* 81, pp. 629–679.

[81] Lundsgaard, E. (1930). "Untersuchungen über Muskelkontraktionen ohne Milchsäurebildung," *Biochemische Zeitschrift* 217, pp. 162–177.

[82] Embden, G., H. J. Deuticke, and G. Kraft (1932). "Über die intermediären

This finding, in minced muscle incubated with hexose diphosphate and fluoride, confirmed that of Ragnar Nilsson (1903–1981), who incubated a dried yeast preparation with hexose diphosphate, acetaldehyde, and fluoride; and isolated phosphoglyceric acid, while the acetaldehyde was reduced to alcohol.[83]

The scheme proposed for glycolysis in muscle by Embden, Deuticke and Kraft was

(1) Fructose-l,6-diphosphate is cleaved to dihydroxyacetone phosphate and D-glyceraldehyde-3-phosphate.
(2) A dismutation of two above trioses to α-glycerophosphate and 3-phosphoglyceric acid.
(3) 3-Phosphoglyceric acid is cleaved to pyruvic acid and inorganic phosphate.
(4) An oxidation-reduction between pyruvic acid and α-glycerophosphate yields lactic acid and D-glyceraldehyde-3-phosphate; the latter product enters reaction 2.

Not only did this scheme fit all the available data and provide for the formation of lactic acid from pyruvic acid instead of methyl glyoxal, but it was confirmed in 1933 by Carl Vincent Smythe (1903–1989), by showing that 50 per cent of DL-glyceraldehyde phosphate is fermented by yeast.[84] This substance had only become available in 1932.[85]

Before 1934 ATP was considered to be a "coenzyme" in glycolysis, but its function was not clear. In that year Lohmann identified the enzyme-catalyzed transfer of a phosphoryl group in the reaction catalyzed by "creatine kinase:"

adenosine triphosphate + creatine ⇆ adenosine diphosphate + creatine phosphate

and Parnas, with his associates Pawel Ostern (1902–1943?) and Thaddeus Mann (1908–1993), made the important discovery that

Vorgänge bei der Glykolyse in der Muskulatur," *Klinische Wochenschrift* 12, pp. 213–215 (213).

[83] Nilsson, R. (1933). "Einige Betrachtungen über den glykolytischen Kohlenhydratabbau," *Biochemische Zeitschrift* 258, pp. 198–206.

[84] Smythe, C. V. and W. Gerischer (1933). "Über die Vergärung der Hexosemonophosphorsäure und 3-Glyceraldehydphosphorsäure," *Biochemische Zeitschrift* 260, pp. 414–416.

[85] Fischer, H. O. L. and E. Baer (1932). "Über die 3-Glycerinaldehydphosphorsäure," *Berichte der deutschen chemischen Gesellschaft* 65, pp. 337–345.

the addition of 3-phosphoglyceric acid to iodoacetate-poisoned minced muscle caused a marked diminution in the production of ammonia; neither fructose-2,6-diphosphate nor pyruvic had this effect. Since the adenylic deaminase discovered by Gerhard Schmidt (1901–1981) did not deaminate ATP, the Parnas group concluded that adenylic acid had been converted to ATP. They stated:

> It follows from this that the resynthesis of creatine phosphate and adenosine triphosphate is not coupled to glycolysis as a whole, but to definite partial processes: and this leads further to the conclusion that this resynthesis does not involve a relationship that may be termed "energetic coupling," but more probably involves the transfer of phosphate residues from molecule to molecule in a reaction similar to the one discovered by Lohmann.[86]

In writing this passage, Parnas may have intended to refer to the conclusion drawn by Meyerhof in 1930 from his measurements of the relation between oxygen and glycogen resynthesis:

> oxidation and resynthesis do not represent a chemically coupled process, for which one can give a stoichiometric equation, but an energetically coupled one.[87]

In 1937 Meyerhof offered the excuse that at the time the substances involved in chemical coupling reactions were unknown.[88]

The considerable contributions of Jacob Karol Parnas to the solution of the problem of the chemical mechanism of muscle glycolysis and alcoholic fermentation place him, along with Meyerhof and Embden, in the front rank of the investigators in this field. He was born in Poland, received his Ph.D. in organic chemistry with Richard Willstätter in Zurich, and he then joined Hofmeister's department of physiological chemistry, where he remained until 1915. Parnas returned to Poland in 1916, was professor of physiological chemistry in Warsaw for three years, and then in Lwów until the German occupation of Poland. He emigrated to Moscow, and headed a laboratory of physiology at the Soviet Academy of Sciences.

---

[86] Lohmann, C. (1934). "Über die enzymatische Spaltung der Kreatinphosphorsäure, zugleich ein Beitrag zum Chemismus der Muskelkontraktion," *Biochemische Zeitschrift* 271, pp. 264–277; Parnas, J. K., P. Ostern, and T. Mann (1934). "Über die Verkettung der chemischen Vorgänge im Muskel," *ibid.*, 272, pp. 64–70 (68–69).

[87] Meyerhof, O. (1930). *Die Chemischen Vorgänge im Muskel*, p. 38. Berlin: Springer.

[88] Meyerhof, O. (1937). "Über die Intermediärvorgänge der enzymatischen Kohlehydratspaltung," *Ergebnisse der Physiologie* 39, pp. 10–75 (41).

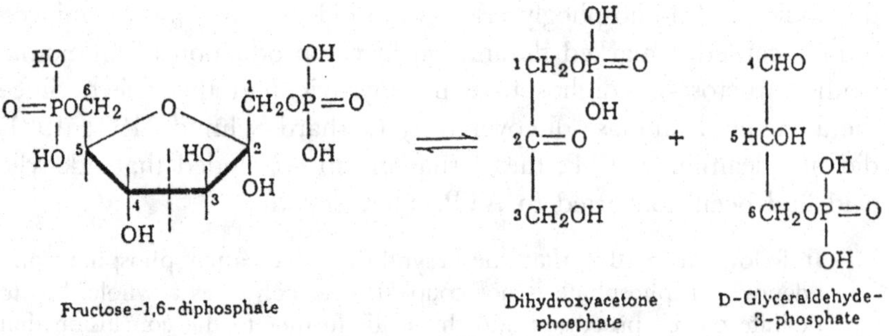

Fructose-1,6-diphosphate $\rightleftharpoons$ Dihydroxyacetone phosphate + D-Glyceraldehyde-3-phosphate

After the appearance of Embden's 1932 scheme, there ensued remarkable ferment in the fermentation community. In 1935 Meyerhof and Lohmann identified the enzyme that cleaves hexose diphosphate to two triose phosphates; they named it "zymohexase." The enzyme was later renamed "aldolase" (after the aldol reaction described by Adolphe Wurtz [1817–1884]) when it was recognized that the immediate products of the cleavage are glyceraldehyde-3-phosphate and dihydroxyacetone phosphate; these are interconverted by a separate enzyme ("triose phosphate isomerase"). The equilibrium in this reversible reaction favors the dihydroxyacetone phosphate, but in the prevailing scheme, only the other triose phosphate is on the direct pathway. In Embden's scheme, 3-phosphoglyceric acid is the precursor of pyruvic acid, and Lohmann showed in 1935 that this conversion involves the catalysis by phosphoglyceromutase of the migration of the phosphoryl group from the 3-position to the 2-position, followed by the dehydration of 2-phosphoglyceric acid to phosphoenolpyruvic acid by the enzyme called enolase. Moreover, in iodoacetate-poisoned muscle, the phosphoryl group of phosphoenolpyruvic acid was transferred to glucose via ATP to form hexose phosphates and pyruvate, and that in alcoholic fermentation the oxidation of glyceraldehyde-3-phosphate to 3-phosphoglyceric acid is balanced by the reduction of acetaldehyde to ethanol.[89]

In 1936, Carl Ferdinand Cori (1896–1984) and Gerty Theresa Cori (1896–1957) added glucose-1-phosphate to the hexose monophosphates; it is formed by the action of the enzyme phosphorylase on

[89] Meyerhof, O. and W. Kiessling (1935). "Über den Hauptweg der Milchsäurebildung in Muskulatur," *Biochemische Zeitschrift* 283, pp. 83–113.

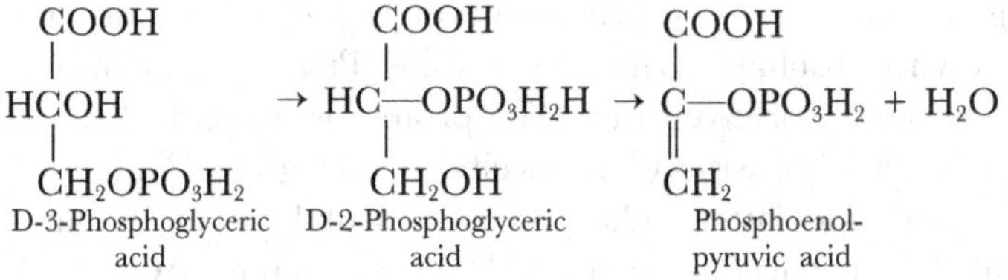

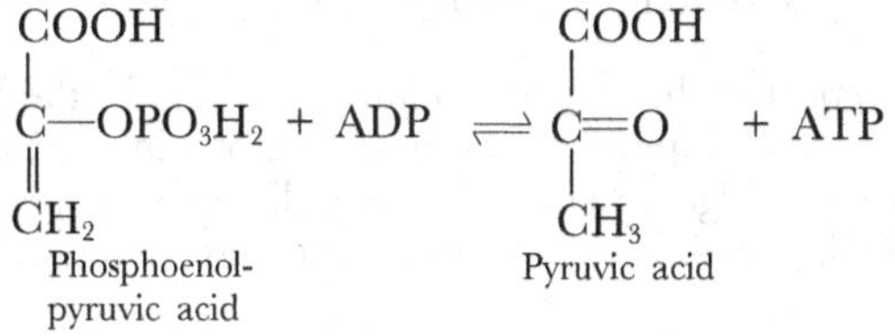

$$\begin{array}{ccccc} \text{COOH} & & \text{COOH} & & \text{COOH} \\ | & & | & & | \\ \text{HCOH} & \rightarrow & \text{HC—OPO}_3\text{H}_2\text{H} & \rightarrow & \text{C—OPO}_3\text{H}_2 + \text{H}_2\text{O} \\ | & & | & & \| \\ \text{CH}_2\text{OPO}_3\text{H}_2 & & \text{CH}_2\text{OH} & & \text{CH}_2 \end{array}$$

D-3-Phosphoglyceric acid — D-2-Phosphoglyceric acid — Phosphoenol-pyruvic acid

$$\begin{array}{ccc} \text{COOH} & & \text{COOH} \\ | & & | \\ \text{C—OPO}_3\text{H}_2 + \text{ADP} & \rightleftharpoons & \text{C=O} \quad + \text{ATP} \\ \| & & | \\ \text{CH}_2 & & \text{CH}_3 \end{array}$$

Phosphoenol-pyruvic acid — Pyruvic acid

glycogen in the presence of inorganic phosphate and is converted to glucose-6-phosphate by a hexose phosphate isomerase.[90] In his obituary notice for Otto Warburg, Hans Krebs wrote:

> By the early 1930s, thanks to the work of Harden, Neuberg, Meyerhof, Embden, the Coris, Parnas, Needham and Lohmann, the enzymes of the intermediary stages of lactic and alcoholic fermentations had been identified and their reactions had been formulated, but not a single one of the enzymes had been obtained in a pure crystalline form. Since the ultimate analysis of the nature of enzyme action depends on the availability of pure substances, the purification of enzymes is of crucial importance.[91]

Apart from the questionable reference to the "early 1930s," this statement reflects a much later concensus. Indeed, in 1928, Warburg had written:

> Since experience teaches that the catalysts of the living substance—the ferments—cannot be separated from their accompanying inactive material, it is appropriate to forego the methods of preparative chemistry, and to study the ferments under the most natural conditions of their activity, in the living cell itself.[92]

---

[90] Cori, C. F. and G. T. Cori (1936). "Mechanism of formation of hexosemonophosphates in muscle and isolation of a new phosphate ester," *Proceedings of the Society for Experimental Biology and Medicine* 34, pp. 702–705.

[91] Krebs, H. A. (1972), p. 651.

[92] Warburg, O. (1928). *Über die Katalytische Wirkung der Lebendigen Substanz*, p. 1. Berlin: Springer.

During the late 1920s, much attention was given to the verdict of the renowned Richard Willstätter (Nobel Prize in Chemistry, 1915) about the nature of invertase: "The protein is no part of the enzyme ... An enzyme consists of a specific active group and a colloidal carrier. With the latter, other substances of high molecular weight are linked in a variable manner."[93] This statement was made at a lecture where James Batcheller Sumner (1887–1955)

> ... recently reported that he had obtained urease in the form of pure crystals which he identifies as those of a globulin. It would perhaps be premature to judge whether the globulin crystals actually are the pure enzyme or whether they only contain the latter in an adsorbed state.[94]

It was not until after 1930, when John Howard Northrop (1891–1987) described not only the isolation of swine pepsin as crystalline protein, but also applied Gibbs's Phase Rule to determine its homogeneity,[95] that news of Sumner's achievement began to arrive in Stockholm. As Northrop's brilliant associate Moses Kunitz (1887–1978) proceeded to crystallize trypsin, chymotrypsin and their zymogens, as well as ribonuclease and deoxyribonuclease, the German biochemists (especially Warburg) caught a glimpse of the future of their discipline. It should also be recalled that in 1934 John Desmond Bernal (1901–1971) and Dorothy Mary Crowfoot (later Hodgkin, 1910–1994) reported X-ray photographs of crystalline pepsin, and after mentioning various ideas about the structure of proteins, they stated:

> At this stage, such ideas are merely speculative, but now that a crystalline protein has been made to give X-ray photographs, it is clear that we have the means of checking them and, by examining the structures of all crystalline proteins, arriving at far more detailed con-

[93] Willstätter, R. (1927). *Problems and Methods in Enzyme Research*, p. 52. Ithaca: Cornell University Press.

[94] *Ibid.*, p. 53. See Sumner, J. B. (1937). "The story of urease," *The Journal of Chemical Education* 14, pp. 255–259; Fruton, J. S. (1977). "Willstätter lectures on enzymes," *Trends in Biochemical Sciences* 2, pp. 210–211.

[95] Northrop, J. H. (1930). "Crystalline pepsin. I: Isolation and tests of purity," *Journal of General Physiology* 13, pp. 739–766; Fruton, J. S. (2002). "A history of pepsin and related enzymes," *The Quarterly Review of Biology* 77, pp. 127–147.

clusions about protein structure than previous physical or chemical methods have been able to give.[96]

The change in Warburg's opinion about the isolation of enzymes and his interest in the crystallization of enzyme proteins was a consequence of the change in his attitude toward the use of dyes such as methylene blue in studies on biological oxidation, about which he had previously been unnecessarily sarcastic. During 1932–1943, with his associates Erwin Negelein, Walter Christian (1907–1959), and Theodor Bücher (1914–1997), Warburg made important contributions to the completion of a coherent scheme of alcoholic fermentation.

In 1929, Warburg visited the United States as a guest of the Rockefeller Foundation, and lectured on 19 October at the Johns Hopkins Hospital.[97] On that occasion, he learned from Eleazar Sebastian Guzman Barron (1898–1957) that, in the presence of glucose, methylene blue greatly increases the normally low rate of oxygen uptake by mammalian erythrocytes. In their published report, Barron and George Argyle Harrop (1890–1945) suggested that

> In favor of the possibility that the principal point at which the methylene blue acts is upon the oxidation of hexose phosphate . . . As to the exact nature of the methylene blue effect little may be said. It is conceivable that it acts as a coenzyme or catalyst, rendering the substrate (hexose phosphate?) more sensitive to the action of molecular oxygen. On the other hand one might consider that methylene blue plays in this system the role ascribed to iron in the oxidations produced by Warburg with his charcoal model.[98]

Warburg undertook the study of the mechanism of the methylene blue effect, and his first conclusion was that it "is nothing but an oxidation by the hemin iron, namely by the iron of methemoglobin,"[99] and that the reaction between methylene blue and glucose is

---

[96] Bernal, J. D. and D. Crowfoot (1934). "X-ray photographs of crystalline pepsin," *Nature* 133, p. 794.

[97] Warburg, O. (1930). "The enzyme problem and biological oxidations," *Bulletin of the Johns Hopkins Hospital* 46, pp. 341–358.

[98] Barron, E. S. G. and G. A. Harrop (1928). "Studies on blood cell metabolism. II: The effect of methylene blue and other dyes upon the glycolysis and lactic acid formation of mammalian and avian erythrocytes," *Journal of Biological Chemistry* 79, pp. 65–87 (85).

[99] Warburg, O., F. Kubowitz, and W. Christian (1930). "Kohlenhydratverbrennung durch Methämoglobin," *Biochemische Zeitung* 221, pp. 494–497 (496).

> . . . a surface reaction. Methylene blue, which is adsorbed on the surfaces of blood cells, forms methemoglobin on the surfaces—that is, at the reaction sites—and therefore a small methemoglobin concentration during methylene blue catalysis to cause a large oxidative effect.[100]

Thus, despite the fact that Barron and Harrop had found no inhibition by cyanide, Warburg concluded that here also "there is a heavy metal catalysis that closely resembles normal catalytic actions of the living substance."[101] Shortly afterwards, however, he and Christian found that although cytolysis of the erythrocytes abolished their ability to oxidize glucose in presence of methylene blue, glucose-6-phosphate was readily oxidized by such cell-free suspensions. After removal of the cell debris by centrifuging the suspension, they fractionated the constituents of the resulting fluid, they were able to conclude that

> the reaction in the blood cells between methemoglobin and hexose monophosphate or between methylene blue and hexose monophosphate occurs by the cooperation of at least two substances, of which we name one "ferment" and the other "coferment."[102]

This finding that the oxidation of glucose-6-phosphate by methylene blue only required a heat-labile nondialyzable "ferment" and a heat-stable dialyzable "coferment" marks a decisive change in Warburg's research strategy. Instead of iron-charcoal models of the *Atmungsferment*, he now dealt with the chemical structure and catalytic function of cozymase, and with the isolation, in crystalline form, of glycolytic enzymes. To the preparative skills needed for these tasks, he brought his experience and apparatus for ultraviolet spectrophotometry.

In 1932, Warburg and Christian isolated from yeast a yellow-red protein which they named "oxygen-transporting ferment"; its pigment was decolorized in the presence of a reducing system composed of glucose-6-phosphate, the "coferment," and an additional "ferment" they found in yeast. In the absence of oxygen, the reduced "leuco" pigment was oxidized by methylene blue. They concluded, therefore, that

---

[100] Warburg, O., F. Kubowitz, and W. Christian (1930). "Über die katalytische Wirkung von Methylenblau in lebenden Zellen," *Biochemische Zeitschrift* 227, pp. 245–271 (270).

[101] *Ibid.*, p. 271.

[102] Warburg, O. and W. Christian (1931). "Über Aktivierung der Robisonschen Hexosemonophosphorsäure in roten Blutzellen und die Gewinnung aktivierender Fermentlösungen," *Biochemische Zeitschrift* 242, pp. 206–227 (215).

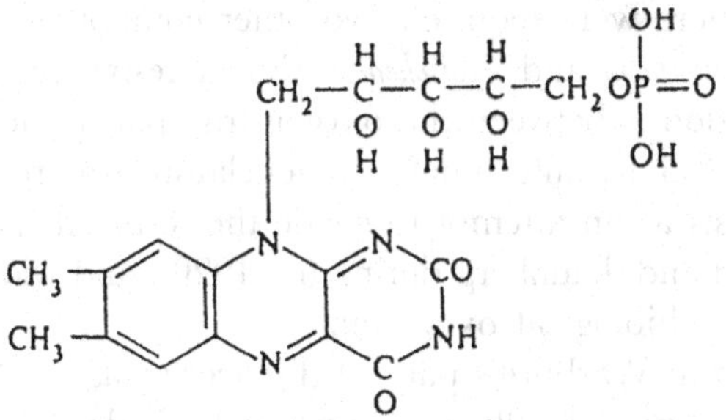

Riboflavin phosphate (flavin mononucleotide)

> The yellow ferment is therefore not only an oxygen-transporting ferment but also a ferment of "oxygen-less respiration" . . . It is probable that in life, the yellow ferment does not transfer molecular, but "bound" oxygen. Probably, in life, it is not an oxygen-transporting ferment but an oxidation-reduction ferment.[103]

The yellow protein was dissociated reversibly by Hugo Theorell (1903–1982) to yield the pigment,[104] whose chemical structure was quickly established by Paul Karrer (1889–1971) and Richard Kuhn (1900–1967) and was named *riboflavin-5'-phosphate* or *flavin mononucleotide* (FMN). As a former associate of Willstätter, Kuhn felt it appropriate to state that

> R. Willstätter thought that an enzyme consists of a solid support and an active group. The explanation that O. Warburg and H. Theorell gave for the structure of the yellow enzyme illustrates exactly this conception.[105]

On the contrary, subsequent work in Warburg's laboratory gave striking evidence for the role of the protein in effecting the catalysis and determining the substrate specificity of the chemical reaction.[106]

We saw earlier that in the aerobic oxidation of glucose-6-phosphate by cytolyzed red cells, in addition to what turned out to be

---

[103] Warburg, O. and W, Christian (1933). "Über das gelbe Ferment und seine Wirkungen," *Biochemische Zeitschrift* 266, pp. 377–411 (377).
[104] Theorell, H. (1935). "Das gelbe Oxydationsferment," *Biochemische Zeitschrift* 278, pp. 263–290.
[105] Kuhn, R. (1935). "Sur les flavines," *Bulletin de la Société de Chimie Biologique* 17, pp. 905–926 (921).
[106] Fruton, J. S. (1972). *Molecules and Life*, pp. 337–338. New York: Wiley.

a flavoprotein there were required two other components; they were denoted *Zwischenferment* and *Zwischen-co-Ferment* respectively, "because their area of action is between the oxygen-transporting ferments and the substrates."[107] This unfortunate nomenclature was recognized by many biochemists as an attempt to evade the term "dehydrogenase" used by Wieland and Thunberg during the 1920s, and widely adopted in writings about biological oxidation.

The next step in Warburg's path of discovery was the finding that the co-ferment from red cells was composed of adenine, two ribose units, three phosphate groups, and a "base I" which he isolated as a picrolonate. It was clear that the coferment from red blood cells was closely related to the cozymase isolated from yeast by Euler during the 1920s; adenine, two ribose units and two phosphate groups were found on hydrolysis.[108] Warburg isolated the free "base I" and sent it for microanalysis to his friend Walter Schoeller (1880–1965) at the Schering company. They had both received their Ph.D. degrees in 1906 for work done in Emil Fischer's institute. Shortly afterward, Warburg and Christian announced that

> Mr. Walter Schoeller has called to our attention that the composition and melting point of Base I agree with those of nicotinic acid amide. A comparison of the two substances showed that they are identical.[109]

Nicotinic acid (pyridine-3-carboxylic acid) had been known since 1870 as an oxidation product of the plant alkaloid nicotine and had been isolated from rice in 1912 by Umetaro Suzuki (1874–1943).

The above announcement was followed by a remarkable paper by Warburg, Christian and Alfred Griese (1918–1943), in which it is stated that

> The pyridine component of the co-ferment is its active group, because the catalytic action of the co-ferment depends on the alternation of the oxidation state of the pyridine part.[110]

---

[107] Warburg, O. and W. Christian (1933), p. 394.

[108] Euler, H. von (1936). "Cozymase," *Ergebnisse der Physiologie* 38, pp. 1–30.

[109] Warburg, O. and W. Christian (1935). "Das Co-Fermentproblem," *Biochemische Zeitschrift* 275, p. 464.

[110] Warburg, O., W. Christian, and A. Griese (1935). "Die Wirkungsgruppe des Co-Ferments aus roten Blutzellen," *Biochemische Zeitschrift* 279, pp. 143–144 (144).

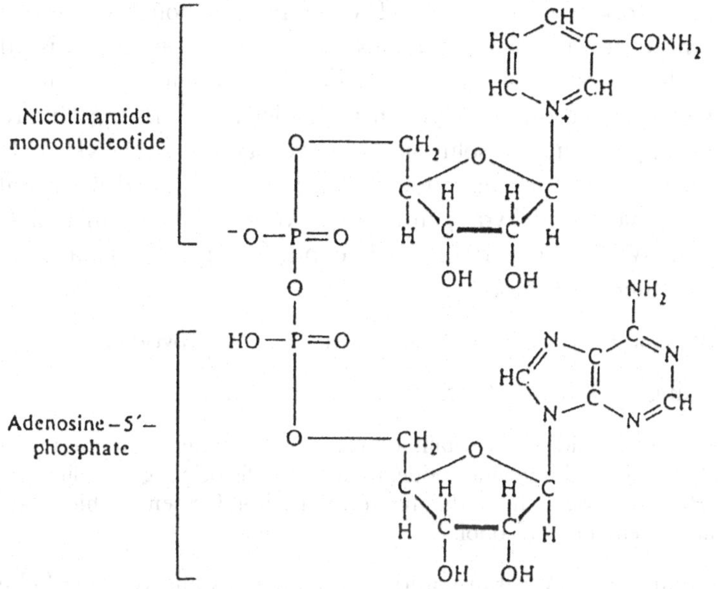

Nicotinamide mononucleotide

Adenosine-5′-phosphate

In 1936, Warburg and Christian isolated from red cells what they called the "fermentation co-ferment" which had two phosphate groups and named it "diphosphopyridine nucleotide" (DPN); the other co-ferment was named "triphosphopyridine nucleotide" (TPN). By 1936, Euler had found nicotinamide in his yeast cozymase, and reiterated his view that cozymase (denoted Cohydrase I) is a general cofactor in dehydrogenase catalyzed reactions. The triphosphate coferment was called Cohydrase II. What Euler called *Apohydrasen*, Warburg called *Gärungs-Zwischenfermente*, and necessary for the reduction of the "fermentation co-ferment." Since about 1960, the terms DPN and TPN have been replaced by NAD (nicotinamide adenine dinucleotide) and NADP (nicotinamide adenine dinucleotide phosphate).

On reduction of the pyridine ring, a new band appears at 340 nm in the ultraviolet absorption spectrum of DPN. Warburg used this property to develop rapid quantitative assays for the DPN-dependent enzymes, and to use this spectrophotometric method for their purification. After 1945, when reliable photoelectric quartz spectrophotometers became available commercially, this method largely replaced the use of the Warburg manometric apparatus and variants of the Thunberg methylene blue technique in studies on dehydrogenases. Warburg referred to the reduced form as *Dihydropyridin*, and later workers denoted it CoH_2 or $DPNH_2$, although it was evident from the chemistry of its reactions that only one hydrogen is added to the 4-position of the ring in a process involving the transfer of two electron; the other hydrogen atom derived from the "hydrogen donor" appears in the solution as a hydrogen ion ($DPNH + H^+$).

The first of the fermentation enzymes to be isolated in crystalline form was alcohol dehydrogenase from yeast by Negelein and Hans Joachim Wulff (1910–1942).[111] The purified protein catalyzed the reaction they wrote as:

Alcohol + Pyridine ⇆ Acetaldehyde + Dihydropyridine

According to Warburg's terminology, they had isolated

> ... this colloid as a crystalline protein. The protein combines with the diphosphopyridine nucleotide to form a dissociating pyridino-protein (*Pyridinoproteid*), the reducing fermentation ferment, which reduces acetaldehyde to alcohol.[112]

Two years later, Warburg and Christian crystallized from Lebedev yeast juice what they called the "oxidizing fermentation ferment" which catalyzes the oxidation of glyceraldehyde-3-phosphate, in the presence of DPN and inorganic phosphate, to 1,3-diphosphoglyceric acid. They pointed out that the reaction which they wrote as:

3-Phosphoglycerinaldehyd + Pyridinnucleotid + H_2O = 3-Phosphoglycerinsäure + Dihydropyridinnucleotid

could not be effected by dialyzed cell extracts, and thus there was an unexplained process in the oxidation of carbohydrates in fermentation.[113]

In 1937, Dorothy Moyle Needham (1896–1987) and Raman Kochukrishna Pillai (1906–1946) had shown that the oxidation of triose phosphate to phosphoglyceric acid in muscle is coupled to the phosphorylation of adenosine diphosphate (ADP) to form ATP and that this coupling is abolished by arsenate, whereas the oxidation-

[111] Negelein, E. and H. J. Wulff (1937). "Diphosphopyridinproteid: Alkohol, Acetaldehyd," *Biochemische Zeitschrift* 293, pp. 351–389.

[112] Warburg, O. and W. Christian (1939). "Isolierung und Krystallisation des Proteins des oxydierenden Gärungsferments," *Biochemische Zeitschrift* 303, pp. 40–68 (40).

[113] *Ibid.*, p. 41.

reduction process is unaffected.[114] Because of its chemical similarity to phosphate Harden and Young had tested the effect of arsenate on the rate of fermentation by yeast juice and found a rate acceleration which they attributed to the rapid hydrolysis of the arsenate analogue of hexose diphosphate. The work of Needham and Pillai suggested instead that arsenate affected the coupled formation of ATP, not only in the case of hexose phosphates, but also for the triose phosphates, with 1,3-diphosphoglyceric acid as an intermediate in the formation of 3-phosphoglyceric acid. Negelein and Brömel succeeded in isolating this labile compound (they called it "R-acid").[115] In aqueous solution it undergoes rapid hydrolysis to 3-phosphoglyceric acid, and in the presence of arsenate the corresponding 1-arseno-3-phosphoglyceric acid is formed and hydrolyzed in a similar manner. The difference lay in the fact that, in the presence of ADP, there is a readily reversible transfer of the 1-phosphoryl group of 1,3-diphosphoglyceric acid to form ATP, whereas the arseno compound only undergoes hydrolysis.

In the 1939 paper on the "protein of the oxidating fermentation ferment" Warburg and Christian noted that its purification through crystallization was necessary because

> In all previous experiments on the oxidation reaction of fermentation the oxidizing fermentation ferment (among others) was contaminated with hexokinase and isomerase. In all previous experiments it did not matter whether the substrate was hexose diphosphate, dihydroxyacetone phosphate, or Fischer ester [3-glyceraldehyde phosphate], there was always an equilibrium of these three substances, and consequently they were all equally reactive. Under these circumstances it was an open question which of the three substances was the substrate of the oxidizing fermentation ferment.[116]

Warburg and Christian also reported that

> Th. Bücher has found in Lebedev juice a specific protein which effects the reaction of the end product of the physiological oxidation reaction and adenosine diphosphate:

---

[114] Needham, D. M. and R. K. Pillai (1937). "The coupling of oxidoreductions and dismutations with esterification of phosphate in muscle," *Biochemical Journal* 31, pp. 1837–1851.

[115] Negelein, E. and H. Brömel (1939). "R-Diphosphoglycerinsäure, ihre Wirkungen und Eigenschaften," *Biochemische Zeitschrift* 303, pp. 132–144.

[116] O. Warburg and W. Christian (1939), p. 45.

1,3-Diphosphoglyceric acid + Adenosine diphosphate ⇆ 3-Phosphoglyceric acid + Adenosine triphosphate.[117]

This important discovery by Theodor Bücher (1914–1997) opened a new chapter in the history of biochemistry in providing a well-defined route for the coupling of an oxidation to the synthesis of ATP from ADP as a model in the study of the energy relationships in biological systems.[118] Because of the outbreak of World War II, the detailed report of the crystallization and properties of the "phosphate-transfer fermentation ferment" did not appear until 1947.[119] Those of the Warburg group drawn into military service included Bücher, Brömel, Griese, and Wulff; of these only Bücher survived the hostilities. During the war, Warburg and Christian isolated and crystallized the fermentation ferments enolase and zymohexase (later named aldolase).[120] Some years after the war, Bücher headed a group in Hamburg who published a paper (in honor of Warburg's seventieth birthday) in which they described the crystallization of aldolase, 3-glyceraldehyde phosphate dehydrogenase, lactic acid dehydrogenase, 3-glycerophosphate dehydrogenase, and pyruvate kinase from rabbit muscle in a single series of operations.[121]

Because of the elegance and significance of Warburg's experiments on fermentation, they had a decisive impact on biochemical research during the 1930s and immediately after World War II. His idiosyncratic nomenclature and his views about the role of the protein components of the pyridine nucleotide-dependent reactions, however, were sources of confusion. For example, David Ezra Green (1910–1983) et al. stated:

---

[117] *Ibid.*, p. 47.

[118] Kalckar, H. M. (1941). "The nature of energetic coupling in biological syntheses," *Chemical Reviews* 28, pp. 71–178.

[119] Bücher, T. (1947). "Über ein phosphatübertragendes Gärungsferment," *Biochimica et Biophysica Acta* 1, pp. 292–314.

[120] O. Warburg and W. Christian (1942). "Isolierung und Krystallisation des Gährungsferments enolase," *Biochemische Zeitschrift* 310, pp. 384–421; (1943). "Isolierung und Krystallisation des Gärungsferments zymohexase," *ibid.* 314, pp. 149–176.

[121] Beisenherz, G., H. J. Boltze, Th. Bücher, R. Czok, K. H. Garbade, E. Meyer-Arendt, and G. Pfleiderer (1953). "Diphosphofructose-Aldolase, Phosphoglyceraldehyde-Dehydrogenase, Milchsäure-Dehydrogenase, Glycerophosphate-Dehydrogenase, und Pyruvat-Kinase aus Kaninchenmuskulatur in einem Arbeitsgang," *Zeitschrift für Naturforschung* 8b, pp. 555–577.

> The concept of "*Zwischenferment*" introduced by Warburg implies that the coenzyme combines with the dehydrogenase to form the catalytically active complex. What is ordinarily referred to as a dehydrogenase is considered by Warburg to be merely a highly specific protein with no catalytic properties apart from its prosthetic group – the coenzyme. Euler and his school have accepted this view but they prefer to call the active complex the "holodehydrase."[122]

Malcolm Dixon (1899–1985) and Leon Zerfas (1897–1978) concluded that "the coenzymes are to be regarded as the prosthetic groups of the dehydrogenases, and it is suggested that the conception of 'pyridine-protein' is misleading and should be abandoned."[123] Moreover, as Parnas noted:

> In biochemical reactions two components participate: in higher concentration and smaller turnover, the coenzymes which act as hydrogen and phosphate transfer agents (acceptors and donors), and the true enzymes, the simple or complicated proteins. We must consider Negelein's dehydrogenase [*Dehydrase*] which hydrogenates from Robinson ester or alcohol 20,000 molecules of coenzyme per minute as no different than the hydrolytic digestive enzymes: up to the point that here the reaction requires the interaction of at least three molecular species. In the seemingly coenzyme-free enzymatic reactions water appears to act as a coenzyme.[124]

These opinions were widely adopted and are reflected in the names of the enzymes in the accompanying "Embden-Meyerhof-Parnas" (EMP) scheme of the pathway in the fermentation by the yeast *Saccharomtces cerevisiae* of glucose to ethanol and CO_2, as presented in a biochemistry textbook published in 1958. Variants of this scheme (which does not show the coupling reactions) have appeared in more up-to-date textbooks, with the replacement of the symbol DPN by NAD. The scheme includes the finding that Harden and Young's "cozymase" also contained a coenzyme for the carboxylase which converts pyruvic acid to acetaldehyde. This co-carboxylase was found by Ernst Auhagen (b. 1904) in 1932, and its structure (shown below)

---

[122] Green, D. E., J. G. Dewan, and L. F. Leloir (1937). "The P-hydroxybutyric dehydrogenase of animal tissues," *Biochemical Journal* 31, pp. 934–949 (948).

[123] Dixon, M. and L. G. Zerfas (1940). "The role of coenzymes in dehydrogenase systems," *Biochemical Journal* 34, pp. 371–391 (391).

[124] Parnas, J. K. (1938). "Über die enzymatischen Phosphorylierungen in der alkoholischen Gärung und in der Muskelglykogenolyse," *Enzymologia* 5, pp. 166–184 (174).

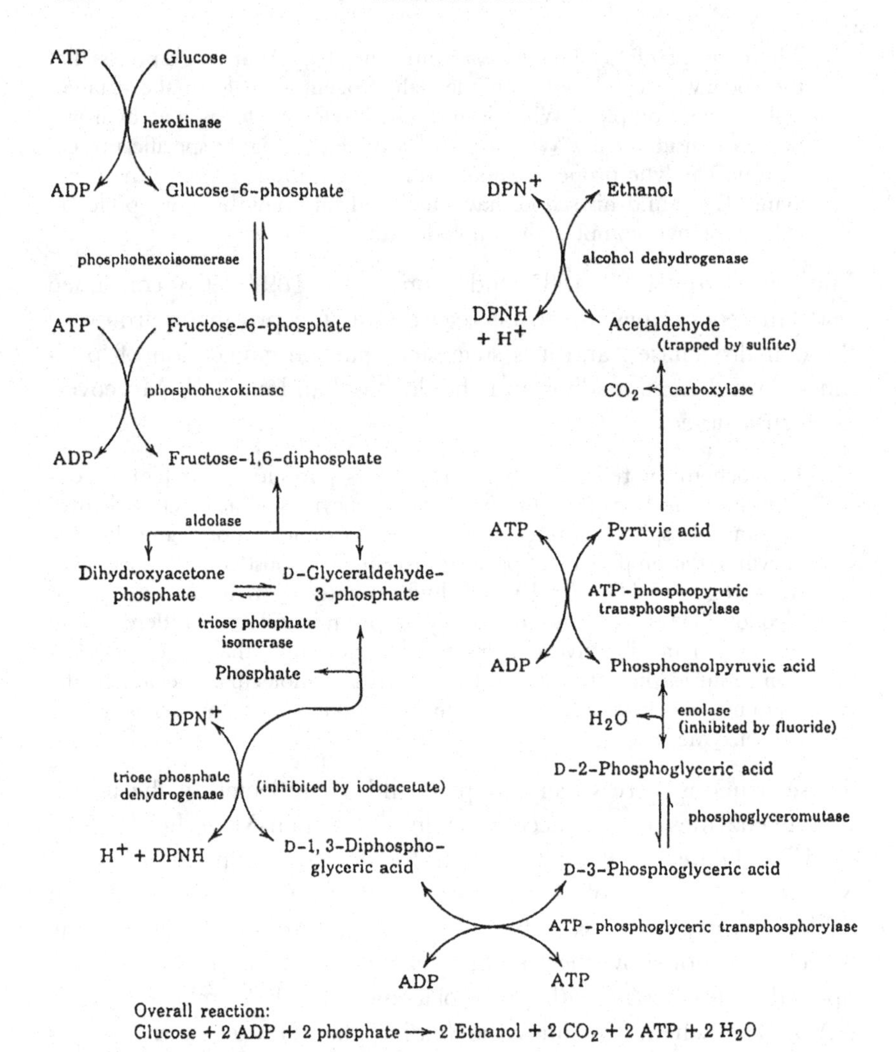

Pathway of Anaerobic Breakdown of Glucose to Ethanol and Carbon Dioxide in Yeast

determined by Karl Lohmann and Philipp Schuster (b. 1908) in 1937.[125]

In a review article written shortly before he was obliged to leave Germany, Meyerhof emphasized the requirement that

[125] Lohmann, K. and P. Schuster (1937). "Untersuchungen über die Cocarboxylase," *Biochemische Zeitschrift* 294, pp. 188–214.

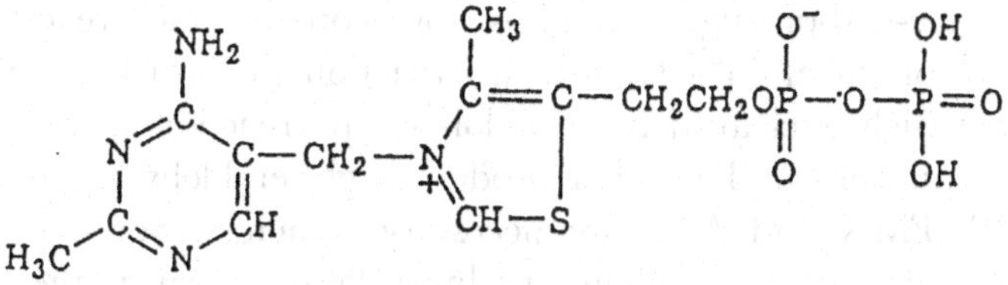

Thiamine pyrophosphate

> In order to consider the individual reactions identified in an isolated enzyme system as intermediate reactions in the total process, it is only acceptable if the starting substrate, in our case glucose or glycogen, can be converted stepwise stoichiometrically into the postulated intermediates and if the rate of each partial reaction in the same enzyme system is at least as great as the rate of the total process.[126]

Some cell-free extracts of dried yeast, however, did not ferment hexose diphosphate. This problem was resolved by Meyerhof after he settled in Philadelphia, by showing that for hexose diphosphate to ferment, sufficient ADP must be made available by the action of an ATP-ase, a labile enzyme readily inactivated in the standard preparation of dried yeast extract.[127]

Although a great achievement, the EMP pathway was not the complete story. Ten years after the end of World War II, Harland Goff Wood (1907–1991), a new leader in fermentation studies, wrote:

> . . . it has been clearly established that glucose is broken down by certain microorganisms by pathways other than the EMP pathway and the mechanism of a whole new series of alternate pathways is being unveiled.[128]

After 1945, with the use of glucose labeled with radioactive carbon, it was shown that whereas the EMP pathway applies to yeast and mammalian muscle, in mammalian liver and red cells the predominant pathway is one in which a hexose-6-phosphate is converted

[126] Meyerhof, O. (1937). "Über die Intermediärvorgänge der enzymatischen Kohlehydratspaltung," *Ergebnise der Physiologie* 39, pp. 10–75 (18).

[127] Meyerhof, O. (1945). "The origin of the reaction of Harden and Young in cell-free alcoholic fermentation," *Journal of Biological Chemistry* 157, pp. 105–119; (1949). "Further studies of the Harden and Young effect in alcoholic fermentation of yeast preparations," *ibid.* 180, pp. 575–586.

[128] Wood, H. G. (1955). "Significance of alternate pathways in the metabolism of glucose," *Physiological Reviews* 35, pp. 841–859.

into a five-carbon sugar (via phosphogluconate) which reacts another five-carbon sugar in a Cannizzaro reaction to form a seven-carbon sugar, which is cleaved in an aldolase-type reaction to regenerate a hexose phosphate. The other product is glyceraldehyde-3-phosphate. NADP, FMN and ATP are necessary cofactors. Some features of this "hexose monophosphate" pathway (also known as the "pentose shunt") have been found in the glucose metabolism of *Lactobacillus pentosus* and in the fixation of CO_2 in photosynthesis. Another alternative pathway of glucose metabolism (known as the "four carbon cycle") was found in the propionic acid bacteria to involve the fixation by phospho-enolpyruvate of CO_2 to yield oxaloacetate, a key component of the tricarboxylic acid (TCA) cycle formulated in 1937 by Hans Adolf Krebs (1900–1981). In this important aerobic pathway of carbohydrate metabolism, oxaloacetate reacts with acetyl-coenzyme A (derived from pyruvate) to form the six-carbon citrate. The new techniques developed after 1945 also made it possible to determine the chemical structure of the enzyme-substrate compounds in the various pathways of metabolism.[129]

Moreover, great progress was made in industrial fermentations, among them the large-scale production of citric acid by *Aspergillus niger* and of penicillin derivatives by various *Penicillium* strains.[130]

---

[129] Walsh, C. (1979). *Enzymatic Reaction Mechanisms.* San Francisco: Freeman.

[130] Raistrick, H. and A. B. Clark (1919). "On the mechanism of oxalic acid formation by *Aspergillus niger*," pp. 13, pp. 329–344; Hastings, J. H. (1971). "The development of the fermentation industries in Great Britain," *Advances in Applied Microbiology* 14, pp. 1–45; Hobby, G. L. (1985). *Penicillin: Meeting the Challenge.* New Haven: Yale University Press.

CONCLUSION

In the preface to his *Micrographia* (1665), the microscopist and physicist Robert Hooke wrote:

> It is the great prerogative of Mankind above other Creatures that we are not only able to *behold* the works of Nature, or barely to *sustein* (sic) our lives by them, but we have also the Power of *considering, comparing, altering, assisting,* and *improving* them to various uses.[1]

In the case of the ancient agricultural or culinary arts of making wines, beer or ale, and bread or cheese, many imaginative conjectures (hypotheses, theories, guesses) were offered about the intimate processes involved in these fermentations and, beginning with Aristotle, analogies were drawn not only with natural processes in living organisms (embryonic development, respiration, digestion, etc.), but also with the generation of metals in the earth. The alchemy that emerged in about 300 A.D., based on the work of metal workers and extractors of plant dyes, laid stress on imagined "philosophical" mercury and sulfur of metals as having some similarity to the mercury ("quicksilver") and sulfur handled in the workshop. Until the end of the seventeenth century, many chemists attempted to effect the transmutation of a base metal into gold by means of a "ferment" (or "philosopher's stone" or "elixir") composed of very pure "philosophical" mercury and sulfur in proper proportion. These efforts were usually conducted under the patronage of emperors, kings, or dukes, and were shrouded in a secretive symbolism comparable in obscure complexity to the present anagrammatic language of the "biomedical" sciences. As in the latter case, the language of alchemy is readily translated by a person familiar with the subject matter.[2] Various kinds of "philosophical furnaces" and elaborate distilling apparatus were used in the alchemical work, and much detailed

[1] Hooke, R. (1944). *Micrographia.* Edinburgh: Oliver and Boyd (Alembic Club reprint No. 5).

[2] Newman, W. R. (1996). "Decknamen and pseudo-chemical language. Eirenaeus Philalethes and Carl Jung," *Revue d'Histoire des Sciences* 49, pp. 159–188.

knowledge was gained about the properties of old and new chemical substances. Before the Muslim takeover in Egypt, alchemy had acquired there an esoteric natural philosophy derived from the local religions, and retained this spiritual character in western Europe until about the end of the seventeenth century. The noted historian of medicine Charles Singer (1876–1960) wrote:

> A modern scientist habitually uses analogy as a means of attaining truth and a guide to experiment, but he never adduces an analogy as a proof of his conclusions. In setting forth his results, indeed, he usually emphasizes his inductive proofs, and thus buries deep among the debris of his working hypotheses the memory of the analogical processes that he has used. This was far from the case of the mediaeval natural philosopher. He started out with the idea that the universe was built on a systematic plan, of the broad meaning of which he believed he had the key.[3]

For example, Walter Pagel (1898–1983) wrote about Paracelsus as follows:

> In all fundamental points of his doctrine, religious motives can be recognized: in the employment of analogy like that of macrocosm and microcosm, in his theory of sympathy and antipathy, in his rejection of ancient humoralism in favor of his doctrine of seeds and created entities. To this last doctrine are due: his view of specificity and virtues, his belief in the importance of solid organs and their metabolism, and his conception of the diseases as entities, varying according to their external causes and their seats. These religious fundamentals are obvious in the predominance of spirit and imagination as emanations of the universal logos of things based on the assumption that they, like Christ, represent the mediator between the One and the Many.[4]

The role of spiritual alchemy in fermentation studies before the eighteenth century is important, but so is the slow accumulation, with limited experimental tools, of reliable practical knowledge about chemical substances. The fact that the pious Robert Boyle, whose self-image as a critic of alchemy concealed his identity as an alchemical adept,[5] or that Isaac Newton, who copied numerous alchemical texts,

[3] Singer, C. (1917). "A review of the medical literature of the dark ages, etc." Reprint from the *Proceedings of the Royal Society of Medicine, Section of the History of Medicine* 10, pp. 1–54 (16–17).

[4] Pagel, W. (1935). "Religious motives in the medical biology of the XVIIth century," *Bulletin of the Institute of the History of Medicine* 3, pp. 97–128 (120).

[5] Principe, L. M. (1998) (note 110).

may have sought there support for his religious preferences,[6] does not lessen the appreciation of Boyle's empirical chemical studies or of the impact of Newton's natural philosophy on eighteenth century chemistry. It was their contemporary, the Oxford physician Thomas Willis, who defined fermentation as "an intestine motion of Particles, or the Principles of every Body, either tending to the Perfection of the same body or because of its change into another."[7] This widely accepted definition was the contribution of the so-called Scientific Revolution to the issue of fermentation.

As chemists have long known, analogy is an error-prone tool.[8] Of course, analogy was decisive in efforts to establish the constitution of organic compounds, and in the development of the periodic system, beginning with the work of Johann Wolfgang Döbereiner.[9] However, the nineteenth-century revival of the vitalist concept of fermentation by Cagniard-Latour, Schwann, and Pasteur was contested by Liebig with wrong arguments and by Moritz Traube with better ones. In 1789, Lavoisier assumed on analogical grounds that what had been known as "spirit of sea salt" contained oxygen; this was disproved by Gay-Lussac and Thenard, who also corrected Lavoisier's analytical data in his fermentation experiment. Other examples are the proposal by the Austrian physicist Leopold Pfaundler (1839–1920) who found an analogy between Charles Darwin's theory of biological evolution and the origin of chemical species,[10] and Emil Fischer's demonstration that his earlier assumption of a structural relationship of arabinose to xylose was analogous to that of glucose and gulose was incorrect.[11]

---

[6] Dobbs, B. T. J. (2000). "Newton as the final cause and first mover" in: *Rethinking the Scientific Revolution*, M. J. Osler (ed.), pp. 25–39. Cambridge University Press.

[7] Dobbs, B. T. J. (1991) (note 5), p. 50.

[8] Snelders, H. A. M. (1994). "Analogie in der chemischen Vergangenheit: Irrwege und Wegweiser," *Schriftenreihe für Geschichte der Naturwissenschaften, Technik und Medizin* N.S. 2, pp. 65–75. See also Farber, E. (1950). "Chemical discoveries by means of analogies," *Isis* 41, pp. 20–26.

[9] Döbereiner, J. W. (1829). "Versuch zu einer Gruppierung der elementaren Stoffe nach ihrer Analogie," *Annalen der Physik und Chemie* 15, pp. 301–307. An English translation is in Leicester, H. M. and H. S. Klickstein (1952). *A Source Book in Chemistry*, pp. 268–272. New York: McGraw-Hill.

[10] Snelders, H. A. M. (1977). "Dissociation, Darwinism, and entropy," *Janus* 64, pp. 51–75.

[11] Fischer, E. (1894). "Synthesen in der Zuckergruppe II," *Berichte der deutschen chemischen Gesellschaft* 26, pp. 3189–3232.

Where does the study of fermentation fit into the history of chemistry? In his *Traité Élémentaire de Chemie*, Lavoisier called vinous fermentation "one of the most striking and extraordinary [operations] of those presented to us by chemistry." In reporting a perfect balance sheet, he used the conversion of sugar to alcohol and carbonic acid gas to confirm anew the principle of the conservation of mass, and to offer the hypothesis that a portion of the sugar is oxidized to carbonic acid and that the rest of the sugar is reduced to alcohol. It is a measure of the transformation of chemistry in the nineteenth century and after World War I that a coherent pathway of carbon in alcoholic fermentation became available during the 1930s. The process turned out to involve the successive catalytic action of twelve specific enzymes, and the role of each enzyme could be established only after it had been purified by crystallization. Steps in the pathway were recognized as analogous to the dismutation reaction of aldehydes discovered by Cannizzaro in 1853 and the aldol condensation discovered in 1872 by Adolphe Wurtz. Emil Fischer's synthetic work on sugars provided an example to be emulated in the synthesis of glyceraldehyde-3-phosphate, whose fermentation by yeast helped to establish the validity of the EMP pathway.

In addition to the decisive importance of organic chemical methods, the elucidation of the pathways of microbial metabolism required the availability of pure cultures of the organisms under study. As was noted earlier in this book, the latter requirement was also essential in the development of industrial fermentation methods. Moreover, during the twentieth century, the effectiveness of both laboratory and large-scale fermenters, as well as the control of such factors as pH, were greatly improved.[12]

---

[12] McNeil, B. and L. M. Harvey (1990). *Fermentation: A Practical Approach.* Oxford University Press; Vogel, H. C. and C. L. Todaro (1997). *Fermentation and Biochemical Engineering Handbook.* Westwood, NJ.: Noyes Publications.

BIBLIOGRAPHY

[Anonymous] (1839). "Das enträthelte Geheimnis der geistigen Gährung," *Annalen der Chemie* 29, pp. 100–104.

Ahrens, F. B. (1902). "Das Gährungsproblem," *Sammlung chemischer und chemisch-technischer Vorträge* 7, pp. 445–494.

Anstey, P. R. (2002), "Boyle on seminal principles," *Studies in History and Philosophy of Biological and Biomedical Sciences* 33, pp. 597–630.

Aristotle (1952). *Meteorologica* (translated by H. D. P. Lee). Cambridge, Mass.: Harvard University Press.

—— (1965). *Historia Animalium* (translated by A. L. Peck). Cambridge, Mass.: Harvard University Press.

Astier, C. B. (1813). "Expériences faites sur le sirop et le sucre de raisin," *Annales de Chimie* 87, pp. 271–285.

Astruc, J. (1711). *Mémoire sur la cause de la Digestion des Alimens.* Montpellier: Honoré Pech.

Bacon, F. (1620). *Novum Organum.* London. .

Bacon, R. (1992). *The Mirror of Alchemy*, S. J. Linden (ed.). New York: Garland.

Baeyer, A. (1870). "Ueber die Wasserentziehung und ihre Bedeutung für das Pflanzenleben und die Gährung," *Berichte der deutschen chemischen Gesellschaft* 3, pp. 63–75.

Bailey, C. H. (1941). "A translation of Beccari's lecture 'Concerning Grain' (1728)," *Cereal Chemistry* 18, pp. 555–561.

Baron, J. H. (1979). "The discovery of gastric acid," *Gastroenterology* 76, pp. 1056–1064.

Barron, E. S. G. and G. A. Harrop (1928). "Studies on blood cell metabolism. II: The effect of methylene blue and other dyes upon the glycolysis and lactic acid formation of mammalian and avian erythrocytes," *Journal of Biological Chemistry* 79, pp. 65–87.

Battelli, F. and L, Stern (1910). "Die Aldehydrase in den Tiergeweben," *Biochemische Zeitschrift* 29, pp. 130–151.

Beaumont, W. (1833). *Experiments and Observations on Gastric Juice and the Physiology of Digestion.* Plattsburg: Allen.

Béchamp. A. (1855). "De l'influence que l'eau pur ou chargée de sels exerce à froid sur le sucre de canne," *Comptes Rendus* 40, pp. 44–47.

Beguin, J. (1624). *Les Elements de Chymie.* 3rd ed. Geneva: Jean Celerier.

Beisenherz, G., H. J. Boltze, Th. Bücher, R. Czok, K. H. Garbade, E. Meyer-Arendt, and G. Pfleiderer (1953). "Diphosphofructose-Aldolase, Phosphoglyceraldehyde-Dehydrogenase, Milchsäure-Dehydrogenase, Glycerophosphate-Dehydrogenase, und Pyruvat-Kinase aus Kaninchenmuskulatur in einem Arbeitsgang," *Zeitschrift für Naturforschung* 8b, pp. 555–577.

Berry, A. J. (1960), *Henry Cavendish.* London: Hutchinson.

Berger, J. (2000), "Atomismus und 'vernunftige chemische Erfahrung': Grundlage der chemischen Materietheorie Georg Ernst Stahls," *Nova Acta Leopoldina* 30, pp. 125–143.

Bernal, J. D. and D. Crowfoot (1934). "X-ray photographs of crystalline pepsin," *Nature* 133, p. 794.

Bernard, C. (1878–1879). *Leçons sur les Phénomènes de la Vie Communs aux Animaux aux Végétaux.* Vol. 1. Paris: Baillière.

—— and Barreswil, C. L. (1845). "Recherches expérimentales sur les phénomènes chimiques de la digestion," *Compt. Rend.* 21, pp. 88–89.

Bernouilli, J. (1997). *Dissertations on the Mechanics of Effervescence and Fermentation and on the Mechanics of the Movement of Muscles*. Philadelphia: American Philosophical Society.
Berthelot, M. (1860). "Sur la fermentation glucosique du sucre de canne," *Comptes Rendus* 50, pp. 980–984.
—— (1860). *Chimie Organique Fondée sur la Synthèse*. Vol. 2. Paris: Mallet-Bachelier.
Berzelius, J. J. (1836). "Einige Ideen über bei der Bildung organischer Verbindungen in der lebenden Natur wirksame, aber bisher nicht bemerkte Kraft," *Jahres-Berichte* 15, pp. 237–245.
—— (1840). *Lehrbuch der Chemie*. Third edition, vol. 9. Dresden: Arnold.
Biringuccio, V. (1540). *Pirotechnia*. Venice (English translation by C. S. Smith and M. T. Gnudi [1943]. Cambridge, Mass.: M.I.T. Press).
Bitting, A. W. (1937). *Appertizing or the Art of Canning: Its History and Development*. San Francisco: The Trade Pressroom.
Björk, R. (2001). "Inside the Nobel Committee on Medicine: Prize competition procedures 1901–1950 and the case of Carl Neuberg," *Minerva* 39, pp. 393–408.
Black, J. (1803). *Lectures on the Elements of Chemistry*, J. Robison (ed.). Edinburgh: Longman & Rees London and Creech Edinburgh.
—— (1944). *Experiments upon Magnesia Alba, Quicklime, and some other Alcaline Substances*. Edinburgh: Oliver and Boyd (Alembic Club Reprint No. 1).
Boas, M. (1956). "Acid and alkali in seventeenth century chemistry," *Archives Internatonales d'Histoire des Sciences* 34, pp. 13–22.
—— and A. R. Hall (1958). "Newton's chemical experiments," *Archives Internationales d'Histoire des Sciences* 11, pp. 113–152.
Boerhaave, H. (1735). *Elements of Chemistry* (translated by T. Dallowe). Vol. 2. London: Pemberton et al.
Böhm, W. (1963). "John Mayow and his contemporaries," *Ambix* 11, pp. 105–120.
Bougard, M. (1999). *La Chimie de Nicolas Lemery*. Turnhout: Brepols.
Boyde, T. R. C. (1980). *Foundation Stones of Biochemistry*. Hong Kong: Voile et Aviron.
Boylan, M. (1982). "The digestive and 'circulatory' systems in Aristotle's biology," *Journal of the History of Biology* 15, pp. 89–118.
Boyle, R. (1999). *The Works of Robert Boyle*, M. Hunter and E. B. Davis (eds.), vol. 3. London: Pickering and Chatto.
Breach, E. F. (1961). "Beccari of Bologna, the discoverer of vegetable protein," *Journal of the History of Medicine* 16, pp. 354–373.
Breathnach, C. S. (2000). "Joseph Black (1728–1799): an early adept in quantification and interpretation," *Journal of Scientific Biography* 8, 149–155.
Brock, W. (1997). *Justus von Liebig. The Chemical Gatekeeper*. Cambridge University Press.
Bücher, T. (1947). "Über ein phosphatübertragendes Gärungsferment," *Biochimica et Biophysica Acta* 1, pp. 292–314.
Buchner, E. (1897). "Alkoholische Gährung ohne Hefezellen (Vorlaufige Mitteilung)," *Berichte der deutschen chemischen Gesellschaft* 30, pp. 117–124.
—— and J. Meisenheimer (1904). "Die chemischen Vorgänge bei der alkoholischen Gärung," *Berichte der deutschen chemischen Gesellschaft* 37, pp. 417–428.
——, H. Buchner, and M. Hahn (1903). *Die Zymasegährung*. Munich: Oldenbourg.
Buchner, H. (1897), "Die Bedeutung der activen löslichen Zellprodukte für den Chemismus der Zelle," *Münchener medizinische Wochenschrift* 44, pp. 299–302.
Buchner, R. (1936). "Die politische und geistige Vorstellungswelt Eduard Buchners," *Zeitschrift für bayerische Landesgeschichte* 26, pp. 631–645.
Caignard de la Tour, C. (1838). "Mémoire sur la fermentation vineuse," *Annales de Chimie* 68, pp. 206–222.
Cavendish, H. (1921). *The Scientific Papers*, E. Thorpe (ed.). Vol. 2. Cambridge University Press.
Chang, K. (2002). "Fermentation, phlogiston, and matter theory: Chemistry and

natural philosophy in Georg Ernst Stahl's *Zymotechnia Fundamentalis*," *Early Science and Medicine* 7, pp. 31–64.
Chaptal, J. A. (1796). *Élémens de Chymie.* 3rd ed. Vol. 3. Paris: Deterville.
Charlton. W. (1659). *Two Discourses. I. Concerning the different Wits of Men. II. Of the Mysterie of Vintners.* London: Willaim Whitwood.
Chick, H. et al. (1971). *War on Disease. A History of the Lister Institute.* London: Andre Deutsch.
Clark, W. M. (1925). "Recent studies on reversible oxidation-reduction in organic systems," *Chemical Reviews* 2, pp. 127–178.
Clericuzio, A. (1990). "A redefinition of Boyle's chemistry and corpuscular philosophy," *Annals of Science* 47, pp. 561–589.
—— (1996). "Alchimie, philosophie corpusculaire et minéralogie dans la Metallographia de John Webster," *Archives Internationales d'Histoire des Sciences* 49, pp. 287–304.
—— (2000), *Elements, Principles and Corpuscles.* Dordrecht: Kluwer.
Cochin, D. (1880). "Recherches du ferment alcoholique soluble," *Annales de Chimie* [5] 21, pp. 430–432.
Coleby, L. J. M. (1938). *The chemical studies of P. J. Macquer.* London: Allen and Unwin.
Connstein, W. and Lüdecke, F. (1919). "Über Glyceringewinnung durch Gärung," *Berichte der deutschen chemischen Gesellschaft* 52, pp. 1385–1391.
Cori, C. F. (1983). "Embden and the glycolytic pathway," *Trends in Biochemical Sciences* 8, pp. 257–259.
—— and G. T. Cori (1936). "Mechanism of formation of hexosemonophosphates in muscle and isolation of a new phosphate ester," *Proceedings of the Society for Experimental Biology and Medicine* 34, pp. 702–705.
Crisciani, C. (1973). "The conception of alchemy as expressed in the *Pretiosa Margarita Novella* of Petrus Bonus of Ferrara," *Ambix* 20, pp. 165–181.
Crosland, M. (1978). *Gay-Lussac: Scientist and Bourgeois.* Cambridge University Press.
Dakin, H. D. and Dudley, H. W. (1913). "Glyoxylase. III. The distribution of the enzyme and its relation to the pancreas," *Journal of Biological Chemistry* 15, pp. 463–474.
Damstaedter, E. (1922). *Die Alchemie des Geber.* Berlin: Springer.
Daumas, M. (1950). "Les appareils d'experimentation de Lavoisier," *Chymia* 3, pp. 45–62.
—— (1955). *Lavoisier Théoricien et Expérimenteur.* Paris: Presses Universitaires de France.
Davenport, H. W. (1991). "Who was Johann Eberle?" *Gastroenterology International* 4, pp. 39–40.
—— (1992), *A History of Gastric Secretion and Digestion.* New York: Oxford University Press.
Davis, A. B. (1973). *Circulation Physiology and Medical Chemistry in England 1650–1680.* Lawrence, Kansas: Coronado Press.
Debus, A. G. (1969). "Edward Jorden and the fermentation of metals. An iatrochemical study of terrestrial phenomena" in: *Towards a History of Geology*, C. J. Schneer (ed.), pp. 101–121.
—— (1991). *The French Paracelsians.* Cambridge University Press.
—— (2001). *Chemistry and Medical Debate. Van Helmont to Boerhaave.* Canton, Mass.: Science History Publications.
Demachy, J. F. (1766). *Instituts de Chymie*, vol. 1. Paris: Lottin.
Deschamps, J. B. (1840). "De la présure," *Journal de Pharmacie* 26, pp. 412–420.
Deuticke, H. J. (1935). "Gustav Embden," *Ergebnisse der Physiologie* 33, pp. 32–49.
Die Alchemie des Andreas Libavius, pp. 103–104. Weinheim: Verlag Chemie.
Diénert, F. V. (1900). *Sur la fermentation de galactose et sur l'accoutumance des levures à ce sucre.* Sceaux: Charaire.
Dixon, M and L. G. Zerfas (1940). "The role of coenzymes in dehydrogenase systems," *Biochemical Journal* 34, pp. 371–391.

Dobbs, B. J. T. (1975). *The Foundations of Newton's Alchemy or "The Hunting of the Greene Lyon."* Cambridge University Press.
—— (1991). *The Janus Face of Genius. The Role of Alchemy in Newton's Thought.* Cambridge University Press.
—— (2000). "Newton as the final cause and first mover" in: *Rethinking the Scientific Revolution,* M. J. Osler (ed.), pp. 25–39. Cambridge University Press.
Döbereiner, J. W. (1829). "Versuch zu einer Gruppierung der elementaren Stoffe nach ihrer Analogie," *Annalen der Physik und Chemie* 15, pp. 301–307. (English translation in: Leicester, H. M. and H. S. Klickstein (1952). *A Source Book in Chemistry,* pp. 268–272. New York: McGraw-Hill.).
Donovan, A. L. (1975). *Philosophical Chemistry in the Scottish Enlightenment.* Edinburgh University Press.
Dubos, R. (1950). *Louis Pasteur: Free Lance of Science.* Boston: Little Brown.
Dubrunfaut, A. P. (1846). "Note sur quelques phénomènes rotatoires et sur quelques propriétes des sucres," *Annales de Chimie* [3] 18, pp. 99–108.
Duclaux, E. (1896). *Pasteur: Histoire d'un Esprit.* Sceaux: Charaire.
Düring, I. (1966). *Aristoteles,* p. 382. Heidelberg: Carl Winter.
Eberle, J. N. (1834). *Physiologie der Verdauung nach Versuchen auf natürlichen und künstlichen Wege.* Wurzburg: Etlinger.
Effront, J. (1899). *Les Enzymes et leurs Applications.* Paris: Cabre & Naud.
Eggleton, P. and M. G. Eggleton (1927). "The physiological significante of 'phosphagen'," *Journal of Physiology* 63, pp. 155–161.
Embden, G. (1924). "Untersuchungen über den Verlauf der Phosphorsäuren und Milchsäure bei der Muskeltätigkeit," *Klinische Wochenschrift* 3, pp. 1393–1396.
—— and G. Schmidt (1929). "Über Muskeladenylsäure und Hefeadenylsäure," *Zeitschrift für physiologische Chemie* 181, pp. 130–139.
—— and M. Oppenheimer (1912). "Über den Abbau der Brenztraubensäure im Tierkörper," *Biochemische Zeitschrift* 45, pp. 186–206.
—— and M. Zimmermann (1927). "Über die Chemie des Lactacidogens. 5. Mitteilung," *Zeitschrift für physiologische Chemie* 167, pp. 114–136.
——, F. Kalberlah, and H. Engel (1912). "Über Milchsäurebildung im Muskelpresssaft," *Biochemische Zeitschrift* 45, pp. 45–62.
——, H. J. Deuticke, and G. Kraft (1932). "Über die intermediären Vorgänge bei der Glykolyse in der Muskulatur," *Klinische Wochenschrift* 12, pp. 213–215.
Erxleben, P. C. F. (1818). *Ueber Guete und Staerke des Bieres, etc.* Prague: Haase.
Euler, H. von (1936). "Cozymase," *Ergebnisse der Physiologie* 38, pp. 1–30.
Farber, E. (1950). "Chemical discoveries by means of analogies," *Isis* 41, pp. 20–26.
Farrington, B. (1953), "On misunderstanding the philosophy of Francis Bacon" in: *Science Medicine and History,* E. A. Underwood (ed.), vol. 1, pp. 439–454. Oxford University Press.
Fernbach, A. and M. Schoen (1913). "L'acide pyruvique, produit de la vie de la levure," *Comptes Rendus* 157, pp. 1478–1480.
Fichman, M. (1971). "French Stahlism and chemical studies of air," *Ambix* 18, pp. 94–122.
Fischer, E. (1894). "Einfluss der Konfiguration auf die Wirkung der Enzyme," *Berichte der deutschen chemischen Gesellschaft* 27, pp. 2985–2993.
—— (1894). "Synthesen in der Zuckergruppe II," *Berichte der deutschen chemischen Gesellschaft* 26, pp. 3189–3232.
—— (1898). "Bedeutung der Stereochemie für die Physiologie," *Zeitschrift für physiologische Chemie* 26, pp. 60–87.
—— and H. Thierfelder (1894). "Verhaltung der verschieden Zucker gegen reine Hefen," *Berichte der deutschen chemischen Gesellschaft* 27, pp. 2031–2037.
Fischer, H. O. L. and E. Baer (1932). "Über die 3-Glycerinaldehydphosphorsäure," *Berichte der deutschen chemischen Gesellschaft* 65, pp. 337–345.

Fiske, C. H. and Y. Subbarow (1927). "The nature of the 'inorganic phosphate' in involuntary muscle," *Science* 65, pp. 401–403.
—— (1949). "Phosphocreatine," *Journal of Biological Chemistry* 81, pp. 629–679.
Fleisch, A. (1924). "Some oxidation processes of normal and cancer tissue," *Biochemical Journal* 18, pp. 294–311.
Fletcher, W. M. and Hopkins, F. G. (1907). "Lactic acid in amphibian muscle," *Journal of Physiology* 35, pp. 247–309.
Forbes, R. J. (1948). *A Short History of the Art of Distillation.* Leiden: Brill.
—— (1949). "Was Newton an alchemist?" *Chymia* 2, 27–36.
—— (1954). "Chemical, culinary, and cosmetic arts" in: *A History of Technology,* C. Singer et al. (eds.), Vol. 1, pp, 238–298. Oxford: Clarendon Press.
Foster, M. (1924). *Lectures on the History of Physiology during the Sixteenth, Seventeenth, and Eighteenth Centuries.* Cambridge University Press.
Fourcroy, A. F. (1799). "D'un mémoire du cit. Fabroni, sur les fermentations etc." *Annales de Chimie* 31, pp. 299–327.
—— (1806). *Philosophie Chimique.* 3rd ed. Paris: Levrault, Schoell. .
Frank, R. G. (1980). *Harvey and the Oxford Physiologists.* Berkeley: University of California Press.
Fraser, P. M. (1972), *Ptolemaic Alexandria,* Vol. 1. Oxford University Press.
—— (1972), *Ptolemaic Alexandria,* Vol. 2. Oxford University Press.
Freeman, K. (1949). *The Pre-Socratic Philosophers.* Oxford University Press.
Freind, J. (1712). *Chymical Lectures.* London: Bawyer.
Fremy, E. (1875). *Sur la Génération des Ferments.* Paris: Masson.
French, J. (1651). *The Art of Distillation.* London: Cates.
Freudenthal, G. (1995). *Aristotle's Theory of Material Substance. Heat and Pneuma, Form and Soul.* Oxford: Clarendon Press.
Freund, I. (1904). *The Study of Chemical Composition.* Cambridge University Press.
Friedman, H. C. (ed.) (1981). *Enzymes.* Stroudsburg, Pennsylvania: Hutchinson Ross.
Fruton, J. S. (1972). *Molecules and Life.* New York: Wiley.
—— (1977). "Willstätter lectures on enzymes," *Trends in Biochemical Sciences* 2, pp. 210–211.
—— (2002). "A history of pepsin and related enzymes," *The Quarterly Review of Biology* 77, pp. 127–147.
Furley, D. (1989). "The mechanics of *Meteorologica* IV: A prolegomenon to biology" in: *Cosmic Problems,* pp. 132–148. Cambridge University Press.
Gantenbein, L. (1992). *Der Chemiater Angelus Sala 1576–1637.* Zurich: Juris.
Gaudillière, J. P. (1991). "Catalyse enzymatique et oxydations cellulaires. L'oeuvre de Gabriel Bertrand et son heritage" in: *L'Institut Pasteur,* M. Morange (ed.), pp. 118–136. Paris: La Découverte.
Gay-Lussac, J. L. (1810). "Mémoire sur la fermentation," *Annales de Chimie* 76, pp. 243–259.
—— (1815). "Lettre à M. Clément, sur l'analyse de l'alcohol et de l'éther sulfurique, et sur les produits de la fermentation," *Annales de Chimie* 95, pp. 311–318.
Geison, G. L. (1981). "Pasteur on vital versus chemical ferments: A previously unpublished paper on the inversion of sugar," *Isis* 72, pp. 425–445.
—— (1995). *The Private Science of Louis Pasteur.* Princeton University Press.
Gelbart, N. R. (1971). "The intellectual development of Walter Charlton," *Ambix* 18, pp. 149–168.
Gerhardt, C. F. (1856). *Traité de Chimie Organique.* Vol. IV. Paris: Firmin-Didot.
Gillespie, L. J. and T. H. Liu (1931). "The reputed dehydrogenation of hydroquinone by palladium black," *Journal of the American Chemical Society* 53, pp. 3969–3972.
Glauber, J. R. (1689). *The Works* (translated by C. Parke). Part II. London: Thomas Milboum.

Golinski, J. (1988). "The secret life of an alchemist" in: *Let Newton Be!*, pp. 147–167.
Goltz, D., J. Telle, and H. J. Vermeer (1977). *Der Alchemistische Traktat 'von der Multiplikation'.* Wiesbaden: Franz Steiner.
Green, D. E., J. G. Dewan, and L. F. Leloir (1937). "The P-hydroxybutyric dehydrogenase of animal tissues," *Biochemical Journal* 31, pp. 934–949.
Green, J. R. (1901). *The Soluble Ferments and Fermentation.* 2nd ed. Cambridge University Press.
Grmek, M. (1972). "Gerbezius, Marcus". *Dictionary of Scientific Biography*, vol. 5. New York: Scribner's.
Grmek, M. D. (1991). "Louis Pasteur, Claude Bemard et la méthode expérimentale" in: *L'Institut Pasteur*, M. Morange (ed.), pp. 21–44. Paris: La Découverte.
Guerlac, H. (1957). "Joseph Black and Fixed Air. A bicentenary retrospective, with some new or little known material," *Isis* 48, pp. 124–151, 433–456.
Gutina, V. (1976). "Sergey Nikolaevich Vinogradsky," *Dictionary of Scientific Biography* 14, pp. 36–38. New York: Scribners.
Guyton-Morveau, L. B. (1778). *Élémens de Chymie Théorique et Pratique.* Vol. 3. Dijon: Frantin.
Haberling, W. (1929). "Alexander von Suchten," *Zeitschrift des Westpreussischen Geschichtsvereins* 69, pp. 177–228.
Hahn, M. (1908). "Zur Geschichte der Zymaseforschung," *Münchener medizinische Wochenschrift* 55, pp. 515–516.
Hales, S. (1727). *Vegetable Staticks.* London: Innis and Woodward.
Hall, A. R. (1998). "Isaac Newton and the aerial nitre," *Notes and Records Royal Society of London* 52, pp. 51–61.
Hall, M. D. V. (1980). "The role of force or power in Liebig's physiological chemistry," *Medical History* 24, pp. 20–59.
Hall, T. E. (1970). "Descartes' physiological method: position, principles, examples," *Journal of the History of Biology* 3, pp. 53–79.
Halleux, R. (1981). *Les Alchimistes Grecs.* Vol. 1. Paris: Les Belles Lettres.
Hammer-Jensen, I. (1921). *Die älteste Alchymie.* Copenhagen: Høst.
Harries, C. (1917). "Eduard Buchner," *Berichte der deutschen chemischen Gesellschaft* 50, pp. 1843–1876.
Haq, S. N. (1994). *Names, Natures and Things. The Alchemist Jābir ibn Hayyām and his* Kitāb al-Ahjār *(Book of Stones).* Dordrecht: Kluwer.
Harden, A. (1903). "Ueber alkoholische Gährung mit Hefe-Pressstoff (Buchners Zymase) bei Gegenwart von Blutserum," *Berichte der deutschen chemischen Gesellschaft* 36, pp. 715–716.
—— (1923). *Alcoholic Fermentation.* 3rd ed. London: Longmans, Green.
—— and W. J. Young (1905). "The alcoholic ferment of yeast juice," *Proceedings of the Royal Society* 77B, pp. 405–420.
Hastings, J. H. (1971). "The development of the fermentation industries in Great Britain," *Advances in Applied Microbiology* 14, pp. 1–45.
Heimann, P. M. (1973). "'Nature is a perpetual worker': Newton's aether and eighteenth-century natural philosophy," *Ambix* 20, pp. 1–25.
Heinecke, B. (1995). "The mysticism and science of Johann Baptist Van Helmont (1579–1644)," *Ambix* 42, pp. 63–78.
Helmholtz, H. (1843), "Ueber das Wesen der Fäulnis und Gährung," *Arch. Physiol.* 5, pp. 453–462. .
Henri, V. (1903). *Lois Générales de L'Action des Diastases.* Paris: Hermann.
Henry, J. (1988). "Newton, matter, and magic" in: *Let Newton Be!*, J. Fauvel et al. (eds), pp. 127–145. Oxford University Press.
Hershbell, J. P. (1987). "Democritus and the beginnings of Greek alchemy," *Ambix* 34, pp. 5–20.
Heym, G. (1938). "Al-Razi and alchemy," *Ambix* 1, pp. 184–191.

Hill, A. V. (1932). "The revolution in muscle physiology," *Physiological Reviews* 12, pp. 56–67.
Hobby, G. L. (1985). *Penicillin: Meeting the Challenge.* New Haven: Yale University Press.
Hoff, H. E. (1964). "Nicholas of Cusa, Van Helmont, and Boyle: The first experiment of the Renaissance in quantitative biology and medicine," *J. Hist. Med.* 14, pp. 99–117.
Hoff, J. H. van't (1898). "Über die zunehmende Bedeutung der anorganischen Chemie," *Zeitschrift für anorganische Chemie* 18, pp. 1–13.
Holmes, F. L. (1985). *Lavoisier and the Chemistry of Life.* Madison, Wis.: University of Wisconsin Press.
—— (1994). "Lavoisier—The conservation of matter," *Chemical & Engineering News* September 12, pp. 38–45.
—— (1998). *Antoine Lavoisier—The Next Crucial Year.* Princeton University Press.
Holmyard, E. J. (1957). *Alchemy.* Harmondsworth: Penguin.
Hooke, R. (1944). *Micrographia.* Edinburgh: Oliver and Boyd (Alembic Club reprint No. 5).
Hooykaas, R. (1949). "The experimental origin of chemical atomic and molecular theory before Boyle," *Chymia* 2, pp. 65–80.
Hopkins, A. J. (1934). *Alchemy Child of Greek Philosophy.* New York: Columbia University Press.
—— (1938). "A study of the kerotakis process as given by Zosimos and later chemical writers," *Isis* 29, pp. 326–354.
Hopkins, F. G. (1921). "The chemical dynamics of muscle," *Johns Hopkins Hospital Bulletin* 32, pp. 359–367.
—— and C. J. Martin (1942). "Arthur Harden (1865–1940)," *Obituary Notices of Fellows of the Royal Society* 4, pp. 3–14.
Hoppe-Seyler, F. (1876). "Ueber die Processe der Gährungen und ihre Beziehung zum Leben des Organismus," *Pflügers Archiv* 12, pp. 1–17.
—— (1878). "Ueber Gährungsprocesse," *Zeitschrift für physiologische Chemie* 2, pp. 1–28.
Höxtermann, E. (1997). "Oscar Brefeld (1839–1925) and the complementary perspective of chemistry and botany toward alcoholic fermentation in the 1870s" in: *Biology Integrating Scientific Disciplines*, B. Hoppe (ed.), pp. 174–188. Munich: Institut für die Geschichte der Wissenschaft.
Hubicki, W. (1960). "Alexander von Suchten," *Gesnerus* 44, pp. 54–63.
Hughes, S. S. (1978). "Martinus Willem Beijerinck," *Dictionary of Scientific Biography* 15, pp. 13–15.
Hunter, M. (ed.) (1994). *Robert Boyle Reconsidered.* Cambridge University Press.
Irvine, W. (1805). *Essays, Chiefly on Chemical Subjects*, W. Irvine, Jr. (ed.). London: Mawman.
Isler, H. (1964). *Thomas Willis.* Zurich.
Ivanov, L. (1906). "Ueber die Synthese der phospho-organischen Verbindungen in abgetöteten Hefezellen," *Zeitschrift für physiologische Chemie* 50, pp. 281–288.
Jaenicke, L. (2002). "Wer begründete die in-vitro-Enzymologie?" *Chemie in unserer Zeit* 36, pp. 64–65.
Joly, B. (1996). "L'alkahest, dissolvant universel ou quand la théorie rend pensable une pratique impossible," *Revue d'Histoire des Sciences* 49, pp. 306–344.
Jorden, E. (1631). *A Discourse of Naturall Bathes, and Minerall Waters.* London: Thomas Harper.
Jørgensen, B. S. (1964). "Berzelius und die Lebenskraft," *Centaurus* 10, pp. 258–281.
Kalckar, H. M. (1941). "The nature of energetic coupling in biological syntheses," *Chemical Reviews* 28, pp. 71–178.
Karger, J. (1957), "Thomas Erastus (1524–1583), der unversöhliche Gegner des Theophrastus," *Gesnerus* 14, pp. 1–13.

Karpenko, V. (1992). "The chemistry and metallurgy of transmutation," *Ambix* 39, pp. 47–62.
Kästner, I. (1996). "Kein Nobelpreis für Maria Manasseina. Ein Beitrag zur Geschichte der Biochemie" in: *Dilettanten und Wissenschaft*, E. Straus (ed.), pp. 123–134.
Kendall, J. (1952). "The first chemical society, the first chemical journal, and the Chemical Revolution," *Proceedings of the Royal Society of Edinburgh* A63, pp. 346–358, 385–400.
Kerker, M. (1955). "Herman Boerhaave and the development of pneumatic chemistry," *Isis* 46, pp. 36–49.
King, L. S. (1964). "Stahl and Hohann: A study in eighteenth-century animism," *Journal of the History of Medicine* 19, pp. 118–130.
—— (1970).*The Road to Medical Enlightenment 1650–1695*. London: Macdonald.
—— (1975). *Dictionary of Scientific Biography* 12. New York: Scribner's.
Kirchhoff, G. S. C. (1815). "Ueber die Zuckerbildung beim Malzen des Getreides etc.," *Schweiggers Journal der Chemie und Physik* 14, pp. 389–398.
Klöcker, A. (1976). "Emil Christian Hansen" in: *The Carlsberg Laboratory 1876–1976*, pp. 168–189. Copenhagen: Rhodos.
Kohler, R. E. (1971). "The reception of Eduard Buchner's discovery of cell-free fermentation," *Journal of the History of Biology* 5, pp. 327–353.
—— (1973). "The enzyme theory and the origin of biochemistry," *Isis* 64, pp. 181–196.
—— (1974). "The background of Arthur Harden's discovery of cozymase," *Bulletin of the History of Medicine* 48, pp. 22–40.
Kohler, R. E., Jr. (1972). "The origin of Lavoisier's first experiments on combustion," *Isis* 63, pp. 349–355.
Kraus, P. (1986). *Jābir ibn Hayyām*. Paris: Les Belles Lettres.
Krebs, H. A. (1972). "Otto Heinrich Warburg (1883–1970)," *Biographical Memoirs of Fellows of the Royal Society* 18, pp. 629–699.
—— (1972). "The Pasteur effect and the relations between respiration and fermentation," *Essays in Biochemistry* 8, pp. 1–34.
Kuhn, R. (1935). "Sur les flavines," *Bulletin de la Société de Chimie Biologique* 17, pp. 905–926.
Kühne, W. (1878). "Erfahrungen und Bemerkungen über Enzyme und Fermente," *Untersuchungen aus dem physiologischen Institut Heidelberg* 1, pp. 291–324.
Kunckel, J. (1716). *Collegium Physico-Chymicum Experimentale oder Laboratorium Chymicum*. Hamburg: éHeyl. Facsimile reprint (1975). Hildesheim: Georg Olms.
Kützing, F. G. (1837). "Mikroskopische Untersuchungen über die Hefe etc.," *Journal für prakische Chemie* 11, pp. 385–409.
Lavoisier, A. (1774). *Opuscules Physiques et Chimiques*. Paris: Deterville. (Annotated English translation [1776] by Thomas Henry: *Essays Physical and Chemical*. London: Joseph Johnson.).
—— (1789). *Traité Élémentaire de Chimie*. Paris: Cuchet. .
Le Febure, N. (1670). *A Compleat Body of Chemistry*. London: Pulleyn.
Le Grand, H. E. (1973). "A note on fixed air: the universal acid," *Ambix* 20, pp. 88–94.
Lebedev, A. (1912). "Extraction de la Zymase par simple maceration," *Annales de L'Institut Pasteur* 28, 8–37.
Lechartier, G. V. and Bellamy, F. (1875). "De la fermentation des fruits," *Comptes Rendus* 81, pp. 1129–1132. .
Lemery, N. (1701). *Cours de Chymie*. 9th ed. Paris: Delespine.
Leuchs, E. F. (1831). "Wirkung des Speichels auf Stärke," *Annalen der Physik und Chemie* 22, p. 623.
Levey, M. (1959). *Chemistry and Chemical Technology in Ancient Mesopotamia*. Amsterdam: Elsevier.

Lewis, E. (1996). *Alexander of Aphrodisias. On Aristotle Meteorology 4*. London: Duckworth.
Liebig, J. (1839). "Ueber die Erscheinungen der Gährung, Fäulnis und Verwesung und ihre Ursachen," *Annalen der Chemie* 30, pp. 250–287.
—— (1842). "Mitscherlich und die Gährungstheorie," *Annalen der Chemie* 41, pp. 357–358.
—— (1870). "Über die Gährung und die Quelle der Muskelkraft," *Annalen der Chemie und Pharmacie* 153, pp. 1–47, 137–229.
—— and Wöhler, F. (1837). "Über die Bildung des Bittermandelöls," *Annalen der Pharmacie* 22, pp. 1–24.
Lindeboom, G. A. (1968). *Herman Boerhaave. The Man and his Work*. London: Methuen.
Lindsay, J. (1970). *The Origins of Alchemy in Graeco-Roman Egypt*. London: Frederick Muller.
Lipman, T.O. (1967). "Vitalism and reductionism in Liebig's physiological thought," *Isis* 58, pp. 167–185.
Lippmann, E. von (1923). "Die chemischen Kenntnisse von Dioscrorides" in: *Abhandlungen und Vorträge zur Geschichte der Naturwissenschaften*. Vol. 1, pp. 47–73. Berlin: Springer.
Lloyd, G. E. R. (1996). "The master cook" in: *Aristotelian Explorations*, pp. 83–103. Cambridge University Press.
Loeb, J. (1906). *The Dynamics of Living Matter*. New York: Columbia University Press.
Loew, O. (1896). *The Energy of Living Protoplasm*. London: Kegan Paul et al.
Lohmann, K. (1928). "Über die Isolierung verschiedener natürlicher Phosphorsäureverbindungen und die Frage ihrer Einheitlichkeit," *Biochemische Zeitschrift* 194, pp. 306–307.
—— (1934). "Über die enzymatische Spaltung der Kreatinphosphorsäure, zugleich ein Beitrag zum Chemismus der Muskelkontraktion," *Biochemische Zeitschrift* 271, pp. 264–277.
—— (1935). "Konstitution der Adenylpyrophosphorsäure und Adenosindiphosphorsäure," *Biochemische Zeitschrift* 282, pp. 120–123.
—— and P. Schuster (1937). "Untersuchungen über die Cocarboxylase," *Biochemische Zeitschrift* 294, pp. 188–214.
Lopez, R. S. (1953). "Hard times and investment in culture" in: *The Renaissance*, pp. 29–54. New York: Harper & Row.
Lu Gwei-Djen, J. Needham and D. Needham (1972). "The coming of ardent water," *Ambix* 19, pp. 69–112.
Lüdersdorff, F. W. (1846). "Ueber die Natur der Hefe," *Annalen der Physik und Chemie* 76, pp. 408–411.
Lundsgaard, E. (1930). "Untersuchungen über Muskelkontraktionen ohne Milchsäurebildung," *Biochemische Zeitschrift* 217, pp. 162–177.
Macbride, D. (1764). *Experimental Essays on the Following Subjects: I. On the Fermentation of Alimentary Mixtures*. London: A. Miller.
Macquer, P. J. (1777). *Elements of the Theory and Practice of Chemistry*. 5th ed. Edinburgh: Donaldson and Elliot. .
Maddison, R. E. W. (1969). *The Life and Works of the Honourable Robert Boyle F.R.S.* London: Taylor & Francis.
Manasseina, M. (1872), "Beiträge zur Kenntnis der Hefe und zur Lehre von der alkoholischen Gährung" in: *Mikroskopische Untersuchungen*, J. Wiesner (ed.), pp. 116–128. Stuttgart.
—— (1872), "Zur Frage von der alkoholischen Gärung ohne lebende Hefezellen," *Berichte der deutschen chemischen Gesellschaft* 30, pp. 3061–3062.
Manchester, K. L. (2000). "Arthur Harden as unwitting pioneer of metabolic control analysis," *Trends in Biochemical Sciences* 25, pp. 89–92.
—— (2001). "Antoine Béchamp: père de la biologie. Oui ou non?" *Endeavour* 25, pp. 68–73.

Mani. N. (1956). "Das Werk von Friedrich Tiedemann und Leopold Gmelin 'Das Verdauung nach Versuchen' und seine Bedeutung für die Entwicklung der Ernährungslehre in der ersten Hälfte des 19. Jahrhunderts," *Gesnerus* 13, pp. 190–214.
Mayow, J. (1907). *Medico-Physical Works* (Alembic Club Reprint No. 12). Edinburgh: Thin.
McCormmach, R. (1961). "Henry Cavendish: A study of rational empiricism in eighteenth-century natural philosophy," *Isis* 60, pp. 293–306.
McDonald, P. (2001). "Remarks on the context of Helmholtz's 'Ueber das Wesen der Fäulnis und Gährung'," *Science in Context* 14, pp. 493–498.
McGovern, P. E. (2003). *Ancient Wine*. Princeton University Press.
McGuire, J. E. (1967). "Transmutation and immutability: Newton's doctrine of physical qualities," *Ambix* 14, pp. 69–95.
—— (1968). "Force, active principles, and Newton's invisible realm," *Ambix* 15, pp. 154–208.
McNeil, B. and L. M. Harvey (1990). *Fermentation: A Practical Approach*. Oxford University Press.
Merton, S. (1966). "Old and new physiology in Sir Thomas Browne: Digestion and some other functions," *Isis* 57, pp. 249–259.
Metzger, H. (1930). *Newton, Stahl, Boerhaave et la Doctrine Chimique*. Paris: Alcan.
Meyerhof, O. (1918). "Über das Gärungscoferment im Tierkörper," *Zeitschrift für physiologische Chemie* 102, pp. 1–32.
—— (1925). "Über den Zusammenhang der Spaltungsvorgänge mit der Atmung in der Zelle," *Berichte der deutschen chemischen Gesellschaft* 58, pp. 991–1001.
—— (1930). *Die Chemischen Vorgänge im Muskel*. Berlin: Springer.
—— (1937). "Über die Intermediärvorgänge der enzymatischen Kohlehydratspaltung," *Ergebnisse der Physiologie* 39, pp. 10–75.
—— (1945). "The origin of the reaction of Harden and Young in cell-free alcoholic fermentation," *Journal of Biological Chemistry* 157, pp. 105–119.
—— (1949). "Further studies of the Harden and Young effect in alcoholic fermentation of yeast preparations," *Journal of Biological Chemistry* 180, pp. 575–586.
—— and W. Kiessling (1935). "Über den Hauptweg der Milchsäurebildung in Muskulatur," *Biochemische Zeitschrift* 283, pp. 83–113.
Mialhe, B. (1856). *Chimie Appliquée à la Physiologie et à la Thérapeutique*. Paris: Masson.
Michaelis, L. and M. L. Menten (1913). "Zur Kinetik der Invertinwirkung," *Biochemische Zeitschrift* 49, pp. 333–336.
Milas, N. A. (1932). "Auto-oxidation," *Chemical Reviews* 10, pp. 295–364.
Mitscherlich, E. (1834). "Ueber die Aetherbildung," *Annalen der Physik* 31, pp. 273–282.
—— (1843). "Über die Gährung," *Berichte der Akademie der Wissenschaften Berlin*, pp. 35–41.
Müller, J. and Schwann, T. (1836). "Versuche über die künstliche Verdauung des geronnen Eiweisses," *Archiv für Anatomie und Physiologie*, pp. 66–89.
Multhauf, R. P. (1948). "Medical chemistry and 'the Paracelsians'," *Bull. Hist. Med.* 28, pp. 101–126.
—— (1954). "John of Rupescissa and the origin of medical chemistry," *Isis* 45, pp. 359–367.
Muralt, A. von (1952). "Otto Meyerhof," *Ergebnisse der Physiologie* 47, pp. i–xx.
Musculus, F. A. (1876). "Sur le ferment de l'urée," *Comptes Rendus* 82, pp. 333–336.
Nägeli, C. (1878). "Theorie der Gärung," *Abhandlungen der königlichen Akademie der Wissenschaften* 13 (2), pp. 77–205.
Needham, D. M. and R. K. Pillai (1937). "The coupling of oxidoreductions and dismutations with esterification of phosphate in muscle," *Biochemical Journal* 31, pp. 1837–1851.
Negelein, E. and H. Brömel (1939). "R-Diphosphoglycerinsäure, ihre Wirkungen und Eigenschaften," *Biochemische Zeitschrift* 303, pp. 132–144.

—— and H. J. Wulff (1937). "Diphosphopyridinproteid: Alkohol, Acetaldehyd," *Biochemische Zeitschrift* 293, pp. 351–389.
Neubauer, A. (2000). "Die Entdeckung der zellfreien Gärung," *Chemie in unserer Zeit* 34, pp. 126–133.
Neubauer, O. and K, Fromherz (1911). "Über den Abbau der Aminosauren bei der Hefegärung," *Zeitschrift für physiologische Chemie* 70, pp. 326–350.
Neuberg, C. and E. Reinfurth (1919). "Weitere Untersuchungen über die korrelative Bildung von Acetaldehyd und Glycerin bei der Zuckerspaltung und neue Beiträge zur Theorie der alkoholischen Gärung," *Berichte der deutschen chemischen Gesellschaft* 52, pp. 1677–1703.
—— and H. Lustig (1942). "Preparation of active zymase extracts from top yeast," *Archives of Biochemistry* 2, pp. 191–196.
—— and J. Kerb (1914). "Über zuckerfreie Hefegärungen. XIII. Zur Frage der Aldehydbildung bei der Gärung von Hexosen sowie bei der sog. Selbstgärung," *Biochemische Zeitschrift* 58, pp. 158–170.
—— and L. Karczag (1911). "Über zuckerfreie Hefegärungen, IV, Carboxylase, ein neues Enzym der Hefe," *Biochemische Zeitschrift* 36, pp. 68–75.
—— and M. Kobel (1929). "Weiteres über die Vorgänge bei desmolytischen Bildung von Methylglyoxal durch Hefe," *Biochemische Zeitschrift* 210, pp. 466–488.
—— and M. Kobel (1930). "Die Zerlegung von nicht phosphoryliertem Zucker durch Hefe unter Bildung von Glycerin und Brenztraubensäure," *Biochemische Zeitschrift* 229, pp. 446–454.
Neumeister, R. (1897). "Bemerkungen zu Eduard Buchners Mitteilungen," *Berichte der deutschen chemischen Gesellschaft* 30, pp. 2963–2966.
Newman, W. R. (1985). "New light on the identity of 'Geber'," *Sudhoffs Archiv* 69, pp. 76–90.
—— (1989). "Technology and alchemical debate in late Middle Ages," *Isis* 80, pp. 423–445.
—— (1991). *The 'Summa perfectionis' of pseudo-Geber*. Leiden: Brill.
—— (1994). *Gehennical Fire. The Lives of George Starkey, an American Alchemist in the Scientific Revolution*. Cambridge, Mass.: Harvard University Press.
—— (1996). "Decknamen and pseudo-chemical language. Eirenaeus Philalethes and Carl Jung," *Revue d'Histoire des Sciences* 49, pp. 159–188.
—— (1999). "Alchemical symbolism and concealment: The chemical house of Libavius" in: *Architecture of Science*, P. Galison and E. Thompson (eds.), pp. 59–77. Cambridge, Mass.: MIT Press.
—— (2001). "Corpuscular anatomy and the tradition of Aristotle's *Meteorology*, with special reference to Daniel Sennert," *International Studies in the Philosophy of Science* 15, pp. 145–153.
—— (2001). "Experimental corpuscular theory in Aristotelian alchemy" in: *Late Medieval and Early Modern Corpuscular Matter Theory*, C. Lüthy et al. (eds.), pp. 291–329.
—— and L. M. Principe (2002), *Alchemy Tried in the Fire. Starkey, Boyle, and the Fate of Helmontian Chemistry*. University of Chicago Press.
Newton, I. (1730). *Opticks*. Fourth ed. (Facsimile edition, 1931). London: Bell.
—— (1959). *The Correspondence of Sir Isaac Newton*. Vol. 1. Cambridge University Press.
—— [1691/1692] (1961). "De Natura Acidorum" in: *The Correspondente of Isaac Newton*, H. W. Trumbull (ed.), Vol. 3 (1688–1694). Cambridge University Press.
Nilsson, R. (1933). "Einige Betrachtungen über den glykolytischen Kohlenhydratabbau," *Biochemische Zeitschrift* 258, pp. 198–206.
Nonclercq, M. (1982). *Antoine Béchamp*. Paris: Maloine.
Northrop, J. H. (1930). "Crystalline pepsin. I: Isolation and tests of purity," *Journal of General Physiology* 13, pp. 739–766.
Obrist, B. (1986). "Die Alchemie in der mittelalterlichen Gesellschaft" in: *Die Alchemie in europäischen Kultur und Wissenschaftsgeschichte*, C. Meinel (ed.), pp. 33–59.

Oldroyd, D. (1973). "An examination of G. E. Stahl's *Philosophical Principles of Universal Chemistry*," *Ambix* 20, pp. 36–52. .
Oldroyd, D. R. (1974). "Some Neo-Platonic and Stoic influences on mineralogy in the sixteenth and seventeenth centuries," *Ambix* 21, pp. 128–156.
Pagel, W. (1935). "Religious motives in the medical biology of the XVIIth century," *Bulletin of the Institute of the History of Medicine* 3, pp. 97–128.
—— (1958). *Paracelsus*. Basel: Karger.
—— (1961). "The prime matter of Paracelsus," *Ambix* 9, pp. 117–135.
—— (1962). "The 'wild spirit' (gas) of John Baptist Van Helmont (1579–1644)," *Ambix* 10, pp. 1–13.
—— (1982). *Joan Baptista Van Helmont*. Cambridge University Press.
Parascandola, J. and A. J. Ihde (1969). "History of the pneumatic trough," *Isis* 60, pp. 351–361.
Parnas, J. (1910). "Über fermentative Beschleunigung der Cannizzaroschen Aldehydumlagerung durch Gewebssäfte," *Biochemische Zeitschrift* 28, pp. 274–294.
Parnas, J. K. (1938). "Über die enzymatischen Phosphorylierungen in der alkoholischen Gärung und in der Muskelglykogenolyse," *Enzymologia* 5, pp. 166–184.
——, P. Ostern, and T. Mann (1934). "Über die Verkettung der chemischen Vorgänge im Muskel," *Biochemische Zeitschrift* 272, pp. 64–70.
Partington, J. R. (1937), "Albertus Magnus on alchemy," *Ambix* 1, pp. 3–20.
—— (1937), "Trithemus and alchemy," *Ambix* 2, pp. 53–59.
—— (1938). "The chemistry of Al-Razi," *Ambix* 1, pp. 192–196.
—— (1956). "The life and work of John Mayow," *Isis* 47, pp. 218–230, 405–417.
—— (1961). *A History of Chemistry*. Vol. 2. London: Macmillan.
—— (1962). *A History of Chemistry*. Vol. 3. London: Macmillan.
—— and D. McKie (1937–1939). "Historical Studies on the phlogiston theory," *Ambix* 2, pp. 361–404.
—— and D. McKie (1937–1939). "Historical Studies on the phlogiston theory," *Ambix* 3, pp. 1–53, 337–371.
—— and D. McKie (1937–1939). "Historical Studies on the phlogiston theory," *Ambix* 4, pp. 113–149.
Pasteur, L. (1861). "Recherches sur la dissymmétrie moléculaire des produits organiques naturels" in: *Société de Chimique de Paris, Leçons de Chimieprofessées en 1860*, pp. 1–48.
—— (1922). *Oeuvres de Pasteur*. Vol. 2. Paris: Masson.
—— (1922). *Oeuvres de Pasteur*. Vol. 6. Paris: Masson.
Patai, R. (1994). *The Jewish Alchemists*, pp. 60–91. Princeton University Press.
Patterson, T. S. (1937). "Jean Beguin and his *Tyrocinium Chymicum*," *Ambix* 2, pp. 243–298.
Paul, H. W. (1996). *Science, Vine, and Wine in Modern France*. Cambridge University Press.
Payen, A. and Persoz, J. F. (1833). "Mémoire sur la diastase etc.," *Annales de Chemie et de Physique* 53, pp. 73–92. (English translation in: Boyde, T. R. C. (1980). *Foundation Stones of Biochemistry*, pp. 45–58. Hong Kong: Voile et Aviron.).
Peck, A. L. (1953). "The cognate *pneuma*," in: *Science Medicine and History*, E. A. Underwood (ed.). Vol. 1, pp. 111–121. Oxford University Press.
Pelouze, J. and Gélis, A. (1843). "Mémoire sur l'acide butynque," *Comptes Rendus* 16, pp. 1262–1271.
Pernety, A. J. (1758). *Dictionnaire Mytho-Hermétique*. Paris: Bauche.
Perrin, C. E. (1982). "A reluctant catalyst: Joseph Black and the Edinburgh reception of Lavoisier's chemistry," *Ambix* 29, pp. 141–176.
Peters, H. (1916). "Leibniz als Chemiker," *Archiv für die Geschichte der Naturwissenschaften und der Technik* 7, pp. 87–106, 220–234, 275–287.
Petrus Bonus (1894). *The New Pearl of Great Price* (translated by A. E. Waite). London: James Elliott.
Pietsch, E. (1956). *Johann Rudolph Glauber*. Munich: Oldenbourg.

Plantefol, L. (1968). "Le genre du mot enzyme," *Comptes Rendus* 266, pp. 41–46.
Plato (1961). *The Collected Dialogues*. E. Hamilton and H. Caims (eds.), pp. 1181–1182. New York: Pantheon Books.
Plessner, M. (1954). "The place of the *Turba philosophorum* in the development of alchemy," *Isis* 45, pp. 331–338.
—— (1973). "Jabir ibn Hayyam," *Dictionary of Scientific Biography* 7, pp. 39–43. New York: Scribners.
Porto, P. A. (2001). "Michael Sendivogius on nitre and the preparation of the philosophers' stone," *Ambix* 48, pp. 1–16.
Poynter, F. N. L. (1953). "A seventeenth-century controversy. Robert Witty versus William Simpson" in: *Science Medicine and History*, E.A. Underwood (ed.). Vol, 2, pp. 72–81. Oxford University Press.
Priesner, C. (1986). "Johann Thölde und die Schriften des Basilius Valentinus" in: *Die Alchemie in der europäischen Kultur- und Wissenschaftsgeschichte*, C. Meinel (ed.), pp. 107–118. Wiesbaden: Harrassowitz.
—— (1997). "Basilius Valentinus und Labortechnik um 1600," *Berichte zur Wissenschaftsgeschichte* 20, pp. 159–172.
Principe, L. (1987). "'Chemical translation' and the role of impurities in alchemy: Examples from Basil Valentine's *Triumphwagen*," *Ambix* 34, pp. 21–30.
Principe, L. M. (1992). "Robert Boyle's alchemical secrecy: codes, ciphers, and concealments," *Ambix* 39, pp. 63–74.
—— (1998). *The Aspiring Adept. Robert Boyle and his Alchemical Quest*. Princeton University Press.
Quercetanus, J. (1605). *The Practise of Chemicall, and Hermeticall Physicke*. Part 2. London: Creede.
Quevenne, T. A. (1938). "Étude microscopique et chimique du Ferment, suivie d'expériences sur la fermentation alcoolique," *Journal de Pharmacie* [2] 24, p. 295.
Raistrick, H. and A. B. Clark (1919). "On the mechanism of oxalic acid formation by *Aspergillus niger*," pp. 13, pp. 329–344.
Rappaport, R. (1961). "G. F. Rouelle: An eighteenth-century chemist and teacher," *Chymia* 6, pp. 68–101.
—— (1962). "Rouelle and Stahl—The phlogistic revolution in France," *Chymia* 7, pp. 73–102.
Rather, L. J. and J. B. Frerichs (1968). "The Leibniz-Stahl controversy—I. Leibniz' opening objections to the *Theoria medica vera*," *Clio Medica* 3, pp. 21–40.
Read, J. (1939). *Prelude to Chemistry*. 2nd ed. London: Bell.
Reaumur, R. A. F. de (1761). *Sur la Digestion des Oiseaux. Second memoire*. Amsterdam: Schreuder et Mortier.
Reinke, J. (1883). "Die Autoxydation in der lebenden Pflanzenzelle," *Botanische Zeitung* 41, 65–76, pp. 89–103.
Riddle, J. M. and J. A. Mulhallond (1980). "Albert on stones and minerals" in: *Albertus Magnus and the Sciences*, J. A. Weisheipl (ed.), pp. 203–234. Toronto: Pontifical Institute of Mediaeval Studies.
Ripley. G. (1652). "The Compound of Alchymie" in: *Theatrum Chemicum Britannicum*, E. Ashmole (ed.), pp. 107–187 (175). London: Nath. Brooke.
Robison, R. (1922). "A new phosphoric ester produced by the action of yeast juice on hexoses," *Biochemical Journal* 16, pp. 809–824.
de Romo, A. C. (1989). "Tallow and the time capsule: Claude Bernard's discovery of the pancreatic digestion of fat," *History and Philosophy of the Life Sciences* 11, pp. 251–274.
Roux, E. (1898). "La fermentation alcoolique et l'évolution de la microbie," *Revue Scientifique* 27, pp. 833–840.
Rubner, M. (1913). "Die Ernährungsphysiologie der Hefezelle bei der alkoholischen Gärung," *Archiv für Physiologie* Supplement, pp. 1–392.

Ruland, M. [1612] (1964). *A Lexicon of Alchemy* (translated by A. E. Waite). London: John M. Watkins.
Ruska, J. (1934). "Die Alchemie von Avicenna," *Isis* 21, pp. 14–51.
—— (1935). *Das Buch der Alaune und Salze*. Berlin: Verlag Chemie.
Salmon, F. (1960). *Aristotle's System of the Physical World*. Ithaca, New York: Cornell University Press.
Schaeffer, S. (1987). "Godly men and mechanical philosophers: Souls and spirits in Restoration natural philosophy," *Science in Context* 1, pp. 5–85.
Scheele, C. W. (1793). *Sämmtliche Physische und Chemische Werke*. Vol. 2. Berlin: Martin Sändig.
Scheider, W. (1972). "Chemistry and iatrochemistry" in: *Science, Medicine, and Society in the Renaissance*, A. G. Debus (ed.), pp. 141–150. New York: Science History Publications.
Schmidt, C. (1847). "Gährungsversuche," *Annalen der Chemie und Pharmacie* 61, pp. 168–174.
—— (1862). "Zur Geschichte der Gährung," *Annalen der Chemie* 126, pp. 126–128.
Schofield, R. E. (1970). *Mechanism and Materialism. British Natural Philosophy in an Age of Reason*. Princeton University Press.
Schriefers, H. (1970), *Dictionary of Scientific Biography* 2, pp. 560–563. New York: Scribners.
Schwann, T. (1836a). "Ueber das Wesen des Verdauungsprocesses," *Archiv für Anatomie und Physiologie*, pp, 90–138.
—— (1836b). "Ueber das Wesen des Verdauungsprocesses," *Annalen der Pharmacie* 20, pp. 28–34.
—— (1837). "Vorläufige Mitteilung, betreffend Versuche über die Weingährung und Fäulnis," *Annalen der Physik* 41, pp. 184–193.
—— (1839). *Mikroskopische Untersuchungen*. Berlin: Sander.
Scott, E. L. (1970). "The 'Macbridean doctrine' of air: An eighteenth-century explanation of some biochemical processes, including photosynthesis," *Ambix* 17, pp. 43–57.
Sennert, D., N. Culpeper, and A. Cole (1662). *Chymistry Made Easie and Useful* etc. London: Peter Cole.
Sernka, T. J. (1979). "Claude Bernard and the nature of gastric acid," *Perspectives in Biology and Medicine* 22, pp. 523–530.
Sherlock, T. P. (1948). "The chemical work of Paracelsus," *Ambix* 3, pp. 33–63.
Siegfried, R. (1989). "Lavoisier and the conservation of weight," *Bulletin for the History of Chemistry* 5, pp. 18–24.
Siegfried, R. (1989). "Lavoisier and the conservation principle," *Bulletin for the History of Chemistry* 5, pp. 18–24.
Simpson, W. (1675). *Zymologiaphysica*. London: W. Cooper.
Singer, C. (1917). "A review of the medical literature of the dark ages, etc." Reprint from the *Proceedings of the Royal Society of Medicine, Section of the History of Medicine* 10, pp. 1–54.
Slator, A. (1906), "Studies in fermentation. Part I: The chemical dynamics of alcoholic fermentation by yeast," *Journal of the Chemical Society* 89, pp. 128–142.
Smeaton, W. A. *Fourcroy Chemist and Revolutionary*. Cambridge: Heffer.
Smith, P. H. (1994). *The Business of Alchemy. Science and Culture in the Holy Roman Empire*. Princeton University Press.
—— (2000). "Vital spirits, redemption, artisanship, and the new philosophy in early modern Europe" in: *Rethinking the Scientific Revolution*, M. J. Osler (ed.), pp. 119–135. Cambridge University Press.
Smythe, C. V. and W. Gerischer (1933). "Über die Vergärung der Hexosemonophosphorsäure und 3–Glyceraldehydphosphorsäure," *Biochemische Zeitschrift* 260, pp. 414–416.

Snelders, H. A. M. (1977). "Dissociation, Darwinism, and entropy," *Janus* 64, pp. 51–75.
—— (1994). "Analogie in der chemischen Vergangenheit: Irrwege und Wegweiser," *Schriftenreihe für Geschichte der Naturwissenschaften, Technik und Medizin* N.S. 2, pp. 65–75.
Sourkes, T. L. (1955). "Moritz Traube (1826–1894). His contribution to biochemistry," *Journal of the History of Medicine* 10, pp. 379–391.
Spallanzani, L. (1789). *Dissertations relative to the Natural History of Animals and Vegetables.* 2 vols. London: J. Murray. .
Spargo, P. E. and C. A. Pounds (1979). "Newton's 'derangement of the intellect.' New light on an old problem," *Notes and Records of the Royal Society of London* 34, pp. 11–32.
Stapleton, H. E., G. L. Lewis, and F. S. Taylor (1949). "The sayings of Hermes quoted in the *Mā Al-Waraqī* of Ibn Umail," *Ambix* 3, pp. 69–90.
Steele, R. (1929). "Practical chemistry in the twelfth century. Rasis de aluminibus et salibus," *Isis* 12, pp. 10–46.
Stillman, J. M. (1924). *The Story of Early Chemistry.* London: Constable.
Ströker, E. (2000). "Georg Ernst Stahls Beitrag zur Grundlegung der chemischen Wissenschaft," *ibid.*, pp. 145–160.
Sumner, J. B. (1937). "The story of urease," *The Journal of Chemical Education* 14, pp. 255–259.
Szent-Györgyi, A. (1924). "Über den Mechanismus der Succin- und Paraphenylendiamin-oxydation. Ein Beitrag zur Theorie der Zellatmung," *Biochemische Zeitschrift* 150, pp. 195–210.
Tachenius, O. (1690), *Clavis to the Ancient Hippocratical Physick or Medicine.* London: Marshal.
Taylor, F. S. (1930). "A survey of Greek alchemy," *Journal of Hellenic Studies* 50, pp. 109–139.
—— (1953). "The idea of the quintessence" in: *Science, Medicine and History*, E. A. Underwood (ed.), vol. 1, pp. 247–265. London: Oxford University Press.
Teich, M. (1983). "Fermentation theory and practice: the beginnings of pure yeast cultivation and English brewing, 1883–1913," *History of Technology* 8, N. Smith (ed.), pp. 117–133.
—— (1992). *A Documentary History of Biochemistry 1770–1940.* Rutherford: Fairleigh Dickinson University Press.
Temple, D. (1986). "Pasteur's theory of fermentation: a virtual tautology?" *Studies in the History and Philosophy of Science* 17, pp. 487–503.
Thackray, A. (1970). *Atoms and Powers: An Essay on Newtonian Matter-Theory and the Development of Chemistry.* Cambridge, Mass.: Harvard University Press.
Thenard, J. L. (1819). *An Essay on Chemical Analysis* (translated by J. G. Children). London: W. Phillips.
Thenard, L. J. (1803). "Mémoire sur la fermentation vineuse," *Annales de Chimie* 46, pp. 294–320.
Theorell, H. (1935). "Das gelbe Oxydationsferment," *Biochemische Zeitschrift* 278, pp. 263–290.
Thunberg, T. (1920). "Zur Kenntnis des intermediären Stoffwechsel und der dabei wirksamen Enzyme," *Skandinavisches Archiv der Physiologie* 40, pp. 1–91.
Traube, M. (1858). *Theorie der Fermentwirkungen.* Berlin: Dummler.
—— (1874). "Ueber das Verhalten der Alkoholhefe in sauerstoffgasfreien Medien," *Berichte der deutschen chemischen Gesellschaft* 7, pp. 872–887.
Traube, M. (1878). "Die chemische Theorie der Fermentwirkungen und der Chemismus der Respiration," *Berichte der deutschen chemischen Gesellschaft* 11, 1984–1992.
Turpin, P. J. F. (1838). "Mémoire sur la cause et les effets de la fermentation alcoolique et aceteuse," *Comptes Rendus* 7, pp. 369–402.

Van Helmont, J. B. (1662). *Oriatrike* (translated by J. L. M.). London: Lodowick Loyd.

Viano, C. (1996). "Aristote et l'alchimie grecque. La transmutation et le modèle aristotélien entre théorie et pratique," *Revue d'Histoire des Sciences* 49, pp. 189–213.

Vogel, H.C. and C. L. Todaro (1997). *Fermentation and Biochemical Engineering Handbook.* Westwood, NJ.: Noyes Publications.

Volhard, J. (1909). *Justus von Liebig.* Vol. 2. Leipzig: Barth.

Waley Singer, D. (1946). "Alchemical texts bearing the name of Plato," *Ambix* 2, pp. 115–128.

Walsh, C. (1979). *Enzymatic Reaction Mechanisms.* San Francisco: Freeman.

Warburg, O. (1914). "Über die Rolle des Eisens in der Atmung des Seeigeleies nebst Bemerkungen über einige durch Eisen beschleunigte Oxydationen," *Zeitschrift für physiologische Chemie* 92, pp. 231–256.

—— (1923). "Über die Grundlagen der Wielandschen Atmungstheorie," *Biochemische Zeitschrift* 142, pp. 518–523. .

—— (1924). "Über Eisen, den sauerstoffübertragenden Bestandteil des Atmungsferments," *Biochemische Zeitschrift* 152, pp. 479–494.

—— (1928). *Über die Katalytische Wirkung der Lebendigen Substanz.* Berlin: Springer.

—— (1929). "Atmungsferment und Oxydasen," *Biochemische Zeitschrift* 214, pp. 1–3.

—— (1930). "The enzyme problem and biological oxidations," *Bulletin of the Johns Hopkins Hospital* 46, pp. 341–358.

—— and E. Negelein (1929). "Über das Absorptionsspectrum des Atmungsferments," *Biochemische Zeitschrift* 214, pp. 64–100.

—— and W. Christian (1931). "Über Aktivierung der Robisonschen Hexosemonophosphorsäure in roten Blutzellen und die Gewinnung aktivierender Fermentlösungen," *Biochemische Zeitschrift* 242, pp. 206–227.

—— and W. Christian (1933). "Über das gelbe Ferment und seine Wirkungen," *Biochemische Zeitschrift* 266, pp. 377–411.

—— and W. Christian (1935). "Das Co-Fermentproblem," *Biochemische Zeitschrift* 275, p. 464.

—— and W. Christian (1939). "Isolierung und Krystallisation des Proteins des oxydierenden Gärungsferments," *Biochemische Zeitschrift* 303, pp. 40–68.

—— and W. Christian (1942). "Isolierung und Krystallisation des Gährungsferments enolase," *Biochemische Zeitschrift* 310, pp. 384–421.

—— and W. Christian (1943). "Isolierung und Krystallisation des Gärungsferments zymohexase," *Biochemische Zeitschrift* 314, pp. 149–176.

——, F. Kubowitz, and W. Christian (1930). "Kohlenhydratverbrennung durch Methämoglobin," *Biochemische Zeitung* 221, pp. 494–497.

——, F. Kubowitz, and W. Christian (1930). "Über die katalytische Wirkung von Methylenblau in lebenden Zellen," *Biochemische Zeitschrift* 227, pp. 245–271.

——, W. Christian, and A. Griese (1935). "Die Wirkungsgruppe des Co-Ferments aus roten Blutzellen," *Biochemische Zeitschrift* 279, pp. 143–144.

Weber, H. H. (1972). "Otto Meyerhof—Werk und Persönlichkeit" in: *Molecular Energetics and Macromolecular Biochemistry*, H. H. Weber (ed.), pp. 3–13. Berlin: Springer.

Webster, C. (1966). "Water is the ultimate principle of nature: The background to Boyle's Sceptical Chemist," *Ambix* 13, pp. 96–107.

—— (2002). "Paracelsus, Paracelsism, and the secularization of the worldview," *Science in Context* 15, pp. 9–27.

Weeks, A. (1997). *Paracelsus. Speculative Theory and the Crisis of the Early Reformation.* Albany: State University of New York Press.

Weindling, P. (1979). *From Bacteriology to Social Hygiene: The Papers of Martin Hahn.* Oxford: Wellcome Unit for the History of Medicine.

Werner, P. (1997). "Learning from an adversary? Warburg against Wieland," *Historical Studies in the Physical and Biological Sciences* 28, pp. 173–196.
Westfall, R. S. (1980). *Never at Rest. A Biography of Isaac Newton.* Cambridge University Press.
Westrumb, J. F. (1788). *Kleine physikalisch-chemische Abhandlungen*, vol. 2. Leipzig: Johann Gottfried Muller.
Wickens, G. M. (ed.) (1952). *Avicenna: Scientist and Philosopher.* London: Luzac.
Wieland, H. (1913). "Über den Mechanismus der Oxydationsvorgänge," *Berichte der deutschen chemischen Gesellschaft* 46, pp. 3327–3342.
—— (1922). "Über den Mechanismus der Oxydationsvorgänge," *Ergebnisse der Physiologie* 20, pp. 477–518.
Willis, T. (1681). *Of Fermentation.* London: Dring et al.
Willstätter, R.(1927). *Problems and Methods in Enzyme Research.* Ithaca: Cornell University Press.
Wilson, C. A. (1988). "Jabirian numbers, Pythagorean numbers, and Plato's *Timaeus*," *Ambix* 35, pp. 1–13.
Wilson, G. (1700). *A Compleat Course of Chymistry.* London: W. Turner.
Witkop, B. (1992). "Remembering Heinrich Wieland: Portrait of an organic chemist and founder of modern biochemistry," *Medical Research Reviews* 12, pp. 195–274.
Wohl, A. (1907). "Die neueren Ansichten über den chemischen Verlauf der Gärung," *Biochemische Zeitschrift* 5, pp. 45–64.
Wood, H. G. (1955). "Significance of alternate pathways in the metabolism of glucose," *Physiological Reviews* 35, pp. 841–859.
Wróblewski, A. (1901). "Ueber den Buchner'schen Hefepresssaft," *Journal für praktische Chemie* [2] 64, pp. 1–70.
Young, J. T. (1998). "Universal medicines: Johann Glauber in England" in: *Faith, Medical Alchemy and Natural Philosophy*, pp. 183–257.
Zelinska, Z. (1987). "Jakub Karol Parnas," *Acta Physiologica Polonica* 38, pp. 91–99.

INDEX OF PERSONAL NAMES

GENERAL INDEX

Phlogiston, 33, 34, 42

History of Science and Medicine Library

ISSN 1872-0684

1. Fruton, J.S. *Fermentation. Vital or Chemical Process?* 2006.
   ISBN 90 04 15268 7, 978 90 04 15268 7